Leveling Up with SQL

Advanced Techniques
for Transforming Data into Insights

Mark Simon

Apress®

Leveling Up with SQL: Advanced Techniques for Transforming Data into Insights

Mark Simon
Ivanhoe VIC, VIC, Australia

ISBN-13 (pbk): 978-1-4842-9684-4   ISBN-13 (electronic): 978-1-4842-9685-1
https://doi.org/10.1007/978-1-4842-9685-1

Managing Director, Apress Media LLC: Welmoed Spahr
Acquisitions Editor: Smriti Srivastava
Development Editor: Laura Berendson
Editorial Assistant: Mark Powers

Cover designed by eStudioCalamar

Cover image by Jigar Panchal on Unsplash (www.unsplash.com)

Distributed to the book trade worldwide by Springer Science+Business Media New York, 1 New York Plaza, Suite 4600, New York, NY 10004-1562, USA. Phone 1-800-SPRINGER, fax (201) 348-4505, e-mail orders-ny@springer-sbm.com, or visit www.springeronline.com. Apress Media, LLC is a California LLC and the sole member (owner) is Springer Science + Business Media Finance Inc (SSBM Finance Inc). SSBM Finance Inc is a **Delaware** corporation.

For information on translations, please e-mail booktranslations@springernature.com; for reprint, paperback, or audio rights, please e-mail bookpermissions@springernature.com.

Apress titles may be purchased in bulk for academic, corporate, or promotional use. eBook versions and licenses are also available for most titles. For more information, reference our Print and eBook Bulk Sales web page at http://www.apress.com/bulk-sales.

Any source code or other supplementary material referenced by the author in this book is available to readers on GitHub (github.com/apress). For more detailed information, please visit https://www.apress.com/gp/services/source-code.

Paper in this product is recyclable

To Brian. You're part of what I am today.

Table of Contents

About the Author xiii

About the Technical Reviewer xv

Acknowledgments xvii

Introduction xix

Chapter 1: Getting Ready 1

   About the Sample Database 1

   Setting Up 3

      Database Management Software 3

      Database Client 4

      The Sample Database 4

   What You Probably Know Already 5

      Some Philosophical Concepts 6

      Writing SQL 10

      Basic SQL 10

      Data Types 11

      SQL Clauses 12

      Calculating Columns 15

      Joins 17

      Aggregates 18

      Working with Tables 20

      Manipulating Data 21

      Set Operations 21

   Coming Up 22

Chapter 2: Working with Table Design..25

Understanding Normalized Tables 25

Columns Should Be Independent..27

Adding the Towns Table...28

Adding a Foreign Key to the Town 29

Update the Customers Table..30

Remove the Old Address Columns...32

Changing the Town 35

Adding the Country...36

Additional Comments 38

Improving Database Integrity..38

Fixing Issues with a Nullable Column...40

Other Adjustments...45

Adding Indexes..49

Adding an Index to the Books and Authors Tables.........................50

Creating a Unique Index 52

Review 54

Normal Form..54

Multiple Values 55

Altering Tables...55

Views 55

Indexes 56

The Final Product...56

Summary..57

Coming Up..58

Chapter 3: Table Relationships and Joins...59

An Overview of Relationships 60

One-to-Many Relationship 61

Counting One-to-Many Joins..64

The NOT IN Quirk 69

Creating a Books and Authors View..70

One-to-One Relationships 72

    One-to-Maybe Relationships 72

Multiple Values 76

    Many-to-Many Relationships 77

    Joining Many-to-Many Tables 82

    Summarizing Multiple Values 84

    Combining the Joins 86

    Many-to-Many Relationships Happen All the Time 90

Another Many-to-Many Example 90

Inserting into Related Tables 93

    Adding a Book and an Author 94

    Adding a New Sale 98

Review 102

    Types of Relationships 103

    Joining Tables 103

    Views 104

    Inserting into Related Tables 104

Summary 104

Coming Up 104

Chapter 4: Working with Calculated Data 105

Calculation Basics 106

    Using Aliases 108

    Dealing with NULLs 113

    Using Calculations in Other Clauses 118

More Details on Calculations 122

    Casting 122

    Numeric Calculations 127

    String Calculations 132

    Date Operations 139

The CASE Expression 151

    Various Uses of CASE 152

Coalesce Is like a Special Case of CASE..154

Nested CASE Expression 155

Summary..158

Aliases..159

NULLs..159

Casting Types..159

Calculating with Numbers 159

Calculating with Strings 160

Calculating with Dates...160

The CASE Expression 160

Coming Up..161

Chapter 5: Aggregating Data **163**

The Basic Aggregate Functions..163

NULL...166

Understanding Aggregates...166

Aggregating Some of the Values...170

Distinct Values 170

Aggregate Filter...171

Grouping by Calculated Values..173

Grouping with CASE Statements 177

Revisiting the Delivery Status...179

Ordering by Arbitrary Strings 181

Group Concatenation..183

Summarizing the Summary with Grouping Sets...185

Preparing Data for Summarizing 186

Combining Summaries with the UNION Clause 189

Using GROUPING SETS, CUBE, and ROLLUP 195

Histograms, Mean, Mode, and Median...201

Calculating the Mean...203

Generating a Frequency Table 203

Calculating the Mode...205

Calculating the Median 207

The Standard Deviation 208

Summary 209

Basic Aggregate Functions 209

NULLs 210

The Aggregating Process 210

Aggregate Filters 210

GROUP BY 211

Mixing Subtotals 211

Statistics 212

Coming Up 212

Chapter 6: Using Views and Friends 213

Working with Views 214

Creating a View 216

Using ORDER BY in MSSQL 219

Tips for Working with View 219

Table-Valued Functions 221

What Can You Do with a View? 225

Caching Data and Temporary Tables 227

Computed Columns 231

Summary 233

Views 233

Table Valued Functions 233

Temporary Tables 234

Coming Up 234

Chapter 7: Working with Subqueries and Common Table Expressions 235

Correlated and Non-correlated Subqueries 239

Subqueries in the SELECT Clause 243

Subqueries in the WHERE Clause 246

Subqueries with Simple Aggregates 246

Big Spenders 246

Last Orders, Please 249

Duplicated Customers 251

Subqueries in the FROM Clause 252

Nested Subqueries 255

Using WHERE EXISTS (Subquery) 258

WHERE EXISTS with Non-correlated Subqueries 259

WHERE EXISTS with Correlated Subqueries 259

WHERE EXISTS vs. the IN() Expression 260

LATERAL JOINS (a.k.a. CROSS APPLY) and Friends 261

Adding Columns 263

Multiple Columns 265

Working with Common Table Expressions 267

Syntax 268

Using a CTE to Prepare Calculations 269

Summary 272

Correlated and Non-correlated Subqueries 272

The WHERE EXISTS Expression 273

LATERAL JOINS (a.k.a. CROSS APPLY) 273

Common Table Expressions 273

Coming Up 273

Chapter 8: Window Functions 275

Writing Window Functions 276

Simple Aggregate Windows 277

Aggregate Functions 279

Aggregate Window Functions and ORDER BY 284

The Framing Clause 285

Creating a Daily Sales View 287

A Sliding Window 288

Window Function Subtotals 290

PARTITION BY Multiple Columns 294

Ranking Functions 296

  Basic Ranking Functions 297

  Ranking with PARTITION BY.. 300

  Paging Results... 302

Working with ntile 305

  A Workaround for ntile... 307

Working with Previous and Next Rows 309

Summary.. 311

  Window Clauses 311

Coming Up.. 312

Chapter 9: More on Common Table Expressions 313

CTEs As Variables... 313

  Setting Hard-Coded Constants 314

  Deriving Constants 316

Using Aggregates in the CTE... 317

  Finding the Most Recent Sales per Customer 317

  Finding Customers with Duplicate Names 319

CTE Parameter Names 320

Using Multiple Common Table Expressions 321

  Summarizing Duplicate Names with Multiple CTEs 322

Recursive CTEs 325

  Generating a Sequence 328

  Joining a Sequence CTE to Get Missing Values... 331

  Daily Comparison Including Missing Days.. 333

  Traversing a Hierarchy.. 336

Working with Table Literals.. 342

  Using a Table Literal for Testing.. 344

  Using a Table Literal for Sorting 348

  Using a Table Literal As a Lookup.. 351

  Splitting a String.. 353

xi

Summary 364

    Simple CTEs 364

    Parameter Names 365

    Multiple CTEs 365

    Recursive CTEs 365

Coming Up 365

Chapter 10: More Techniques: Triggers, Pivot Tables, and Variables 367

Understanding Triggers 368

    Some Trigger Basics 369

    Preparing the Data to Be Archived 370

    Creating the Trigger 372

    Pros and Cons of Triggers 380

Pivoting Data 381

    Pivoting the Data 382

    Manually Pivoting Data 384

    Using the Pivot Feature (MSSQL, Oracle) 389

Working with SQL Variables 394

    Code Blocks 395

    Updated Code to Add a Sale 396

Review 404

    Triggers 404

    Pivot Tables 405

    SQL Variables 405

Summary 406

Appendix A: Cultural Notes 407

Appendix B: DBMS Differences 411

Appendix C: Using SQL with Python 421

Index 443

About the Author

Mark Simon has been involved in training and education since the beginning of his career. He started as a teacher of mathematics, but quickly pivoted into IT consultancy and training because computers are much easier to work with than high school students. He has worked with and trained in several programming and coding languages and currently focuses mainly on web development and database languages. When not involved in work, you will generally find him listening to or playing music, reading, or just wandering about.

About the Technical Reviewer

 **Aaditya Pokkunuri** is an experienced senior cloud database engineer with a demonstrated history of working in the information technology and services industry with 13 years of experience.

He is skilled in performance tuning, MS SQL Database Server Administration, SSIS, SSRS, PowerBI, and SQL development.

He possesses in-depth knowledge of replication, clustering, SQL Server high availability options, and ITIL processes.

His expertise lies in Windows administration tasks, Active Directory, and Microsoft Azure technologies.

He also has extensive knowledge of MySQL, MariaDB, and MySQL Aurora database engines.

He has expertise in AWS Cloud and is an AWS Solution Architect Associate and AWS Database Specialty.

Aaditya is a strong information technology professional with a Bachelor of Technology in Computer Science and Engineering from Sastra University, Tamil Nadu.

Acknowledgments

The sample data includes information about books and authors from Goodreads (`www.goodreads.com/`), particularly from their lists of classical literature over the past centuries. Additional author information was obtained, of course, from Wikipedia (`www.wikipedia.org/`).

The author makes no guarantees about whether the information was correct or even copied correctly. Certainly, the list of books should not in any way be interpreted as an endorsement or even an indication of personal taste. After all, it's just sample data.

Introduction

In the early 1970s, a new design for managing databases was being developed based on the original work of E. F. Codd. The underlying model was known as the relational model and described a way of collecting data and accessing and manipulating data using mathematical principles.

Over the decade, the SQL language was developed, and, though it doesn't follow the relational model completely, it attempts to make the database accessible using a simple language.

The SQL language has been improved, enhanced, and further developed over the years, and in the late 1980s, the language was developed into a standard of both ANSI (the American National Standards Institute) and ISO (the International Organization for Standardization, and, that's right, it doesn't spell ISO).

The takeaways from this very brief history are

- SQL has been around for some time.

- SQL is based on some solid mathematical principles.

- There is an official standard, even if nobody quite sticks to it.

- SQL is a developing language, and there are new features and new techniques being added all the time.

The second half of the third point is worth stressing. Nobody quite sticks to the SQL standards. There are many reasons for this, some good, some bad. But you'll probably find that the various dialects of SQL are about 80–90% compatible, and the rest we'll fill you in on as we go.

In this book, you'll learn about using SQL to a level which goes beyond the basics. Some things you'll learn about are newer features in SQL; some are older features that you may not have known about. We'll look at a few non-standard features, and we'll also look at using features that you already know about, but in more powerful ways.

This book is not for the raw beginner—we assume you have some knowledge and experience in SQL. If you are a raw beginner, then you will get more from my previous book, *Getting Started with SQL and Databases*;[1] you can then return to this book full of confidence and enthusiasm with a good solid grounding in SQL.

If you have the knowledge and experience, the first chapter will give you a quick overview of the sort of knowledge you should have.

The Sample Database

To work through the exercises, you'll need the following:

- A database server and a suitable database client.

- Permissions to do anything you like on the database. If you've installed the software locally, you probably have all the permissions you need, but if you're doing this on somebody else's system, you need to check.

- The script which produces the sample database.

The first chapter will go into the details of getting your DBMS software and sample database ready. It will also give you an overview of the story behind the sample database.

Notes

While you're writing SQL to work with the data, there's a piece of software at the other end responding to the SQL. That software is referred to generically as a database server, and, more specifically, as a DataBase Management System, or DBMS to its friends. We'll be using that term throughout the book.

The DBMSs we'll be covering are PostgreSQL, MariaDB, MySQL, Microsoft SQL Server, SQLite, and Oracle. We'll assume that you're working with reasonably current versions of the DBMSs.

Chapter 1 will go into more details on setting up your DBMS, as well as downloading and installing the sample database.

Source Code

All source code used in this book can be downloaded from `github.com/apress/ leveling-up-sql`.

---

[1]`https://link.springer.com/book/978148429494`.

CHAPTER 1

Getting Ready

If you're reading this book, you'll already know some SQL, either through previous study or through bitter experience, or, more likely, a little of both. In the process, there may be a few bits that you've missed, or forgotten, or couldn't see the point.

We'll assume that you're comfortable enough with SQL to get the basic things done, which mostly involves fetching data from one or more tables. You may even have manipulated some of that data or even the tables themselves.

We *won't* assume that you consider yourself an expert in all of this. Have a look in the section "What You Probably Know Already" to check the sort of experience we think you already have. If there are some areas you're not completely sure about, don't panic. Each chapter will include some of the background concepts which should take you to the next level.

If all of this is a bit new to you, perhaps we can recommend an introductory book. It's called *Getting Started with SQL and Databases* by Mark Simon, and you can learn more about it at `https://link.springer.com/book/10.1007/978-1-4842-9493-2`.

About the Sample Database

For the sample database, we're going to suppose that we're running an online bookshop: **BookWorks**. In this scenario

- Customers visit the website.

- At some point, customers will have registered with their details.

- They then add one or more copies of one or more books to a shopping cart.

- Hopefully, they then check out and pay.

- BookWorks will then procure the books and ship them to customers at some point.

1

© Mark Simon 2023
M. Simon, *Leveling Up with SQL*, https://doi.org/10.1007/978-1-4842-9685-1_1

To manage all of this, the database tables look something like Figure 1-1.

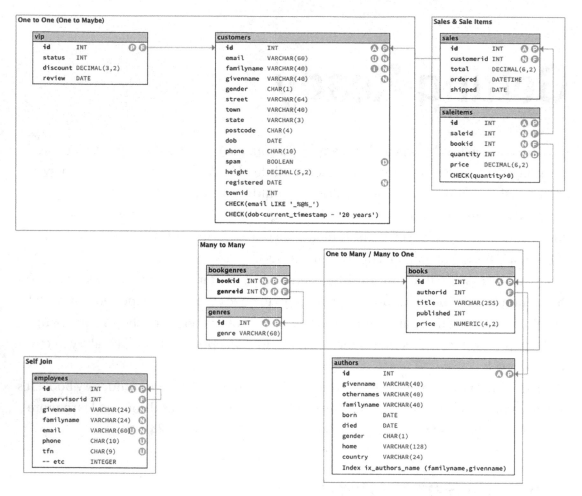

Figure 1-1. *The BookWorks Schema*

In real life, there's more to the story. For example, we haven't included payment or shipping methods, and we haven't included login credentials. There's no stock either, although we'll presume that the books are ordered on demand.

But there's enough in this database for us to work with as we develop and improve our SQL skills.

Setting Up

You can sit in a comfortable chair with a glass of your favorite refreshment and a box of nice chocolates and read this book from cover to cover. However, you'll get more from this book if you join in on the samples.

Database Management Software

First, you'll need access to Database Management Software (DBMS). The five and a half DBMSs we work with in the book are

- PostgreSQL

- MariaDB/MySQL

- Microsoft SQL Server

- SQLite

- Oracle

PostgreSQL, MariaDB/MySQL, and SQLite are all free. Microsoft SQL Server and Oracle are paid products, but have free versions.

MariaDB is a spin-off of MySQL, which is why they're treated together. They are almost identical in features, but you'll find a few places where they're not identical.

If you're using MariaDB/MySQL, we're going to assume that you're running it in **ANSI mode**. It's easily done if you start the session with

```
SET SESSION sql_mode = 'ANSI';
```

You'll probably see this message a few times throughout the book. The Appendix will tell you why.

It's possible—even likely—that you already have the DBMS installed. Just make sure that

- It's a fairly recent version.

  Some of the features you'll learn about aren't available in some older versions of some DBMSs. In particular, watch out for MySQL: you'll need version 8 which was released in 2018 for some of the more sophisticated features.

- You have enough privileges to create a database and to create and modify tables. Most of the book won't require that, but Chapter 2 definitely will.

  At the very least, you'll need to be able to install the sample database.

If you can't make changes to the database, you can still work with most of the book, and you'll just have to nod your head politely as you're reading Chapter 2, in which we make a few changes to the database. You might also have some difficulty in creating views, which we cover in Chapter 6 and in other chapters.

Database Client

You'll also need a database client. All the major DBMS vendors have their own free client, and there are plenty of free and paid third-party alternatives.

The Sample Database

And, of course, you'll need to install the sample database.

The sample database and additional code files for this book are available on GitHub via the book's product page, located at `www.apress.com/ISBN`.

You can also directly download a script by visiting

`www.sample-db.net/`

and clicking a few buttons.

You'll need to do the following:

1. For your DBMS, create your sample database. If you can't think of a better name, bookworks is a fine name. For most DBMSs, you can run

   ```
   CREATE DATABASE bookworks;
   ```

   You'll then need to connect to the database.

2. Using the preceding link, select the options for your DBMS.

   For this sample, you should select the "Book Works" sample (Step 2), as well as the additional Towns and Countries tables (Step 4).

   Download the file. It will come as a ZIP file, so you'll have to unzip it.

3. Using your database, connect to your new database, open the downloaded script file, and run the script.

What You Probably Know Already

... or, A Crash Course in SQL

If you get a sense of déjà vu reading what follows, it's a summary of what you would have learned from my prior Apress book, *Getting Started with SQL and Databases*. You can skip to the next chapter if you're confident with these ideas, but it might be worth going over, to keep them fresh.

In this section, we'll go over the following ideas:

- Some Philosophical Concepts
- Writing SQL
- Basic SQL
- Data Types
- SQL Clauses
- Calculating Columns
- Joins

5

- Aggregates

- Working with Tables

- Manipulating Data

- Set Operations

This is a summary of what you will have encountered in the prior book. Some of these topics will be pushed further in the following chapters.

Some Philosophical Concepts

Some people get the wrong idea of what computers do and, in particular, what's going on in a database. Here, we'll look at clearing this up, as well as clarifying the terminology—what the words mean.

A **database** is a collection of data. Well, obviously, but when we're talking about SQL, we're talking about data which is organized and accessed in a particular way. To begin with, the design of a database follows what is called the **relational model**, which is basically a set of principles about how the data is organized. This model is all about purity and clarity. Each item of data has exactly one place where it belongs and is stored in its purest form. Related items of data are collected together.

Relational database purists won't be hesitant to point out that SQL databases don't follow these principles to the letter or, in some cases, even to the paragraph. Nevertheless, the relational model is the basis of how SQL databases are put together.

Data vs. Information vs. Values

Databases store **data**. That's what the name implies, but it's important to understand that that's not the same as **information**, even if we yield to the temptation to call it that.

Data is neutral. It has no meaning. Your height might be, say, 175 (cm), but the database neither knows nor cares whether that's good or bad. It's just a number, and if it's not correct, it doesn't care about that either.

What the database does care about, however, is whether the data entered follows any rules predefined in the design of the database. That might include the type of data entered or the range of possible values.

Information, however, is something that humans do. We assign it a meaning, and we make judgments. Here, we decide whether the height is what we'd expect or meaningful in some other way.

Why would that be important? Take, for example, your date of birth. Is it possible that it might change?

The short answer is no, you can't (as far as we know) go back and change your date of birth. However, the actual *data* itself can change, such as when it was entered incorrectly, or there's been a change to the calendar (which doesn't, admittedly, happen often).

This affects how a database should be designed: you have to allow for errors, and you have to see what reasonableness checks you might need to add to the definitions. You can't, for example, lock in the date of birth, just because it's not supposed to change.

The other concept is the **value**. Think of the data as a question and the value as the answer. What is your given name (data)? The answer is its value.

That's important, because much of the design of a database is about the data, not the actual values.

For example, a well-constructed database should only store your given name data exactly once. However, the actual value ("Fred," "Wilma," etc.) might well appear with somebody else's data. Values can be repeated, and, if they are, we just regard that as a coincidence. The real giveaway is that you might change the value of one person's given name without being obliged to do the same elsewhere.

To put it simply:

- **Data** is a placeholder. It should never be duplicated elsewhere.

- A **value** is the content of the data. It may be NULL which means that you don't have the value, and it may be duplicated because, well, these things happen.

- **Information** is the meaning you personally put on the database, and the database neither knows nor cares about that. We won't be dealing with information much here.

We may use the term "information" loosely to refer to data, but it's really not the same thing.

Database Tables

SQL databases store data in one or more **tables**. In turn, a table presents the data in **rows** and **columns**. You get the picture in Figure 1-2.

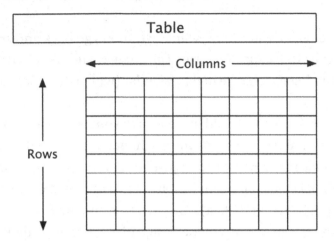

Figure 1-2. *A Database Table*

A row is an instance of the data, such as a book in the books table or a customer in the customers table. Columns are used for details, such as the given name of a customer or the title of a book. Figure 1-3 gives the idea.

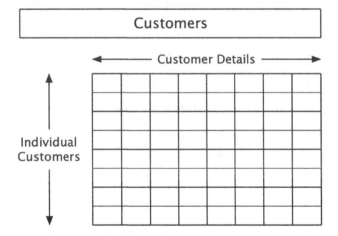

Figure 1-3. *A Customers Table*

Here are some of the important properties of a well-designed table:

- Data is atomic: In each row, each column stores one piece of data only.

- Row order is not significant: You can sort them if you like, but the row order has no real significance.

- Rows are unique: You don't have two rows describing the same thing.

- Rows are independent: Data in one row shouldn't affect any other row.

- Columns are independent of each other: Changing what's in one column shouldn't affect what's in another column.

- Columns are of a single type: You can't mix types in a single column.

- Column names are unique. Obviously.

- Column order is not significant: That's a bit confusing, since obviously column order may be the only clue as to which is which. However, it doesn't matter which order you choose.

One important consequence of this is that columns should never be used to hold multiple values, either singly or in combination. This means

- A single column should not contain multiple values.

- Multiple columns cannot have the same role.

There are a few additional rules, but they are more fine-tuning of the basic principles. SQL uses the term "table" in two overlapping ways:

- Data is stored in the database in a table. The data is accessed as rows and columns; that is, the data is in a table format.

  There is such a thing as a **temporary table**. That's the same as a real table earlier, except that it will self-destruct when you finish with the session.

- Data may also be held fleetingly in a table format without actually being stored.

  You'll get this table data as a result of a join, a common table expression, a view, or even from another SELECT.

When we need to refer to the generated table data, we'll use the term **virtual table** to make the point clear.

Writing SQL

SQL is a simple language which has a few rules and a few recommendations for readability:

- SQL is relaxed about using extra spacing. You should use as much spacing as required to make your SQL more readable.

- Each SQL statement ends with a semicolon (;).

- The SQL language is case insensitive, as are the column names. Table names may be case sensitive, depending on the operating system.

Microsoft SQL is relaxed about the use of semicolons, and many MSSQL developers have got in the bad habit of forgetting about them. However, Microsoft strongly encourages you to use them, and some SQL may not work properly if you get too sloppy. See `https://docs.microsoft.com/en-us/sql/t-sql/language-elements/transact-sql-syntax-conventions-transact-sql#transact-sql-syntax-conventions-transact-sql`.

If you remember to include semicolons, you'll stay out of trouble.

Remember, some parts of the language are flexible, but there is still a strict syntax to be followed.

Basic SQL

The basic statement used to fetch data from a table is the SELECT table. In its simplest form, it looks like this:

```
SELECT ...
FROM ...;
```

- The SELECT statement will select one or more columns of data from a table.

- You can select columns in any order.

- The SELECT * expression is used to select all columns.

- Columns may be calculated.

Calculated columns should be named with an alias; noncalculated columns can also be aliased.

A comment is additional text for the human reader which is ignored by SQL:

- SQL has a standard single-line comment: -- etc

- Most DBMSs also support the non-standard block comment: /* ... */

- Comments can be used to explain something or to act as section headers. They can also be used to disable some code as you might when troubleshooting or testing.

Data Types

Broadly, there are three main data types:

- Numbers

- Strings

- Dates and times

Number literals are represented bare: they do not have any form of quotes.

Numbers are compared in number line order and can be filtered using the basic comparison operators.

String literals are written in single quotes. Some DBMSs also allow double quotes, but double quotes are more correctly used for column names rather than values.

- In some DBMSs and databases, upper and lower case may not match.

- Trailing spaces should be ignored, but aren't always.

Date literals are also in single quotes.

- The preferred date format is ISO8601 (yyyy-mm-dd), though Oracle doesn't like it so much.

- Most DBMSs allow alternative formats, but avoid the ??/??/yyyy format, since it doesn't mean the same thing everywhere.

Dates are compared in historical order.

SQL Clauses

For the most part, we use up to six clauses in a typical SELECT statement. SQL clauses are written in a specific order. However, they are processed in a slightly different order, as in Figure 1-4.

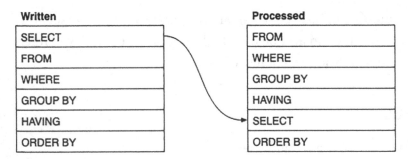

Figure 1-4. *SQL Clause Order*

The important thing to remember is that the SELECT clause is the last to be evaluated before the ORDER BY clause. That means that only the ORDER BY clause can use values and aliases produced in the SELECT clause.[1]

As we'll see later in the book, there are additional clauses which are extensions to the one we have here.

---

[1] SQLite is the exception here. You can indeed use aliases in the other clauses.

Filtering Data with the WHERE Clause

A table can be filtered using the WHERE clause.

When you have a large number of rows, you can filter them using the WHERE clause. The WHERE clause is followed by one or more assertions which evaluate either to true or false, determining whether a particular row is to be included in the result set.

The syntax for the WHERE clause is

```
SELECT columns
FROM table
WHERE conditions;
```

The conditions are one or more assertions, expressions which evaluate to true or not true. If an assertion is not true, it's not necessarily false either. Typically, if the expression involves NULL, the result will be unknown, which is also not true.

- NULL represents a missing value, so testing it is tricky.

- NULLs will always fail a comparison, such as the equality operator (=).

Testing for NULL requires the special expression IS NULL or IS NOT NULL.

Multiple Assertions

You can combine multiple assertions with the logical AND and OR operators. If you combine them, AND takes precedence over OR.

The IN operator will match from a list. It is the equivalent of multiple OR expressions. It can also be used with a subquery which generates a single column of values.

Wildcard Matches

Strings can be compared more loosely using wildcard patterns and the LIKE operator.

- Wildcards include special pattern characters.

- Some DBMSs allow you to use LIKE with non-string data, implicitly converting them to strings for comparison.

- Some DBMSs supplement the standard wildcard characters with additional patterns.

- Some DBMSs support regular expressions, which are more sophisticated than regular wildcard pattern matching.

13

Sorting with the ORDER BY Clause

SQL tables are unordered collections of rows.

- Row order is insignificant and may be unexpected.

- You can sort the results using an ORDER BY clause.

Sorting a table is done using the ORDER BY clause:

```
SELECT columns
FROM table
--  WHERE ...
ORDER BY ...;
```

The ORDER BY clause is both the last to be written and the last to be evaluated.

- Sorting does not change the actual table, just the order of the results for the present query.

- You can sort using original columns or calculated values.

- You can sort using multiple columns, which will effectively group the rows; column order is arbitrary, but will affect how the grouping is effected.

- By default, each sorting column is sorted in increasing (ascending) order. Each sorting column can be qualified by the DESC clause which will sort in decreasing (descending) order. You can also add ASC which changes nothing as it's the default anyway.

- Different DBMSs will have their own approach as to where to place sorted NULLs, but they will all be grouped either at the beginning or the end.

- Data types will affect the sort order.

- Some DBMSs will sort upper and lower case values separately.

Limiting Results

A SELECT statement can also include a limit on the number of rows. This feature has been available unofficially for a long time, but is now an official feature.

The official form is something like

```
SELECT ...
FROM ...
ORDER BY ... OFFSET ... ROWS FETCH FIRST ... ROWS ONLY;
```

This is supported in PostgreSQL, MSSQL, and Oracle.

Many DBMSs still offer their proprietary unofficial limiting clauses. The most common unofficial version is something like

```
SELECT ...
FROM ...
ORDER BY ... LIMIT ... OFFSET ...;
```

This is supported in PostgreSQL (which also supports OFFSET ... FETCH), MariaDB/MySQL, and SQLite.

MSSQL also has a simple TOP clause added to the SELECT clause.

Sorting Strings

Sorting alphabetically is, by and large, meaningless. However, there are techniques to sort strings in a more meaningful order.

Calculating Columns

In SQL, there are three main data types: numbers, strings, and dates. Each data type has its own methods and functions to calculate values:

- For numbers, you can do simple arithmetic and calculate with more complex functions. There are also functions which approximate numbers.

- For dates, you can calculate an age between dates or offset a date. You can also extract various parts of the date.

- For strings, you can concatenate them, change parts of the string, or extract parts of the string.

- For numbers and dates, you can generate a formatted string which gives you a possibly more friendly version.

Calculating with NULLs

Whenever a calculation involves a NULL, it has a catastrophic effect on the result, and the result will normally be NULL.

In some cases, you may be able to substitute a value using `coalesce()` which will replace NULL with a reasonable alternative. Of course, you will need to work out what you mean by "reasonable."

Aliases

Every column should have a distinct name. When you calculate a value, you supply this name as an alias using AS. You can also do this with noncalculated columns to provide a more suitable name.

Aliases and other names should be distinct. They should also follow standard column naming rules, such as not being the same as an SQL keyword and not having special characters.

If, for any reason, a name or an alias needs to break the naming rules, you can always wrap the name in double quotes (`"double quotes"`) or whatever the DBMS supplies as an alternative.

Some DBMSs have an alternative to double quotes, but you should prefer double quotes if possible.

Subqueries

A subquery is an additional SELECT statement used as part of the main query.

A column can also include a value derived from a subquery. This is especially useful if you want to include data from a separate related table. If the subquery involves a value from the main table, it is said to be correlated. Such subqueries can be costly, but are nonetheless a useful technique.

The CASE Expression

You can generate categories using CASE ... END, which tests a value against possible matches and results in one out of a number of alternative values.

Casting a Value

You may be able to change the data type of a value, using `cast()`:

- You can change within a main type to a type with more or less detail.

- You can sometimes change between major types if the value sufficiently resembles the other type.

Sometimes, casting is performed automatically, but sometimes you need to do it yourself.

One case where you might need to cast from a string is when you need a date literal. Since both string and date literals use single quotes, SQL might misinterpret the date for a string.

Views

You can save a `SELECT` statement into the database by creating a view. A view allows you to save a complex statement as a virtual table, which you can use later in a simpler form.

Views are a good way of building a collection of useful statements.

Joins

Very often, you will create a query which involves data from multiple tables. Joins effectively widen tables by attaching corresponding rows from the other tables.

The basic syntax for a join is

```
SELECT columns
FROM table JOIN table;
```

There is an older syntax using the `WHERE` clause, but it's not as useful for most joins.

Although tables are joined pairwise, you can join any number of tables to get results from any related tables.

When joining tables, it is best to distinguish the columns. This is especially important if the tables have column names in common:

- You should fully qualify all column names.

- It is helpful to use table aliases to simplify the names. These aliases can then be used to qualify the columns.

17

The ON Clause

The ON clause is used to describe which rows from one table are joined to which rows from the other, by declaring which columns from each should match.

The most obvious join is from the child table's foreign key to the parent table's primary key. More complex joins are possible.

You can also create *ad hoc* joins which match columns which are not in a fixed relationship.

Join Types

The default join type is the INNER JOIN. The INNER is presumed when no join type is specified:

- An INNER JOIN results only in child rows for which there is a parent. Rows with a NULL foreign key are omitted.

- An OUTER JOIN is an INNER JOIN combined with unmatched rows. There are three types of OUTER JOIN:

  - A LEFT or RIGHT join includes unmatched rows from one of the joined tables.

  - A FULL join includes unmatched rows from both tables.

  - A NATURAL join matches two columns with identical names and doesn't require an ON clause. It is particularly useful in joining one-to-one tables. Not all DBMSs support this.

There is also a CROSS JOIN, which combines every row in one table with every row in the other. It's not generally useful, but can be handy when you cross join with a single row of variables.

Aggregates

Instead of just fetching simple data from the database tables, you can generate various summaries using aggregate queries. Aggregate queries use one or more aggregate functions and imply some groupings of the data.

Aggregate queries effectively transform the data into a secondary summary table. With grand total aggregates, you can only select summaries. You cannot also select non-aggregate values.

The main aggregate functions include

- `count()`, which counts the number of rows or values in a column

- `min()` and `max()` which fetch the first or last of the values in sort order

For numbers, you also have

- `sum()`, `avg()`, and `stdev()` (or `stddev()`) which perform the sum, average, and standard deviation on a column of numbers

When it comes to working with numbers, not all numbers are used in the same way, so not all numbers should be summarized.

For strings, you also have

- `string_agg()`, `group_concat()`, or `listagg()`, depending on the DBMS, which concatenates strings in a column

In all cases, aggregate functions only work with values: they all skip over `NULL`. You can control which values in a column are included:

- You can use `DISTINCT` to count only one instance of each value.

- You can use `CASE ... END` to work as a filter for certain values.

Without a `GROUP BY` clause, or using `GROUP BY ()`, the aggregates are grand totals: you will get one row of summaries.

You can also use `GROUP BY` to generate summaries in multiple groups. Each group is distinct. When you do, you get summaries for each group, as well as additional columns with the group values themselves.

Aggregates are not limited to single tables:

- You can join multiple tables and aggregate the result.

- You can join an aggregate to one or more other tables.

In many cases, it makes sense to work with your aggregates in more than one step. For that, it's convenient to put your first step into a common table expression, which is a virtual table which can be used with the next step.

When grouping your data, sometimes you want to filter some of the groups. This is done with a `HAVING` clause, which you add after the `GROUP BY` clause.

Working with Tables

Tables are created using the CREATE TABLE statement. This statement includes

- Column names
- Data types
- Other table and column properties

A table design can be changed afterward, such as adding triggers or indexes. More serious changes, such as adding or dropping columns, can be effected using ALTER TABLE statements.

Data Types

There are three main types of data:

- Numbers
- Strings
- Dates

There are many variations of the preceding types which make data storage and processing more efficient and help to validate the data values.

There are also additional types such as boolean or binary data, which you won't see so much in a typical database.

Constraints

Constraints define what values are considered valid. Standard constraints include

- NOT NULL
- UNIQUE
- DEFAULT
- Foreign keys (REFERENCES)

You can construct your own additional constraints with the generic CHECK constraint. Here, you add a condition similar to a WHERE clause which defines your own particular validation rule.

Foreign Keys

A foreign key is a reference to another table and is also regarded as a constraint, in that it limits values to those which match the other table.

The foreign key is defined in the child table.

A foreign key also affects any attempt to delete a row from the parent table. By default, the parent row cannot be deleted if there are matching child rows. However, this can be changed to either (a) setting the foreign key to NULL or (b) cascading the delete to all of the children.

Indexes

Since tables are not stored in any particular order, they can be time consuming to search. An optional index can be added for any column you routinely search, which makes searching much quicker.

Manipulating Data

Data manipulation statements are used to add or change data. In addition to the SELECT statement, there are

- INSERT: Add new rows to the table

- UPDATE: Change the data in one or more rows in the table

- DELETE: Delete one or more rows of the table

Like SELECT, the UPDATE and DELETE statements can be qualified with a WHERE clause to determine which rows will be affected.

Unlike SELECT, these have the potential to make a mess of a database, especially since SQL doesn't have an undo.

Set Operations

In SQL, tables are mathematical sets of rows. This means that they contain no duplicates and are unordered. It also means that you can combine tables and virtual tables with set operations.

There are three main set operations:

- UNION combines two or more tables and results in all of the rows, with any duplicates filtered out. If you want to keep the duplicates, you use the UNION ALL clause.

- INTERSECT returns only the rows which appear in all of the participating tables.

- EXCEPT (a.k.a. MINUS in Oracle) returns the rows in the first table which are not also present in the second.

When applying a set operation, there are some rules regarding the columns in each SELECT statement:

- The columns must match in number and type.

- Only the names and aliases from first SELECT are used.

- Only the values are matched, which means that if your various SELECTs change the column order or select different columns, they will be matched if they are compatible.

A SELECT can include any of the standard clauses, such as WHERE and GROUP BY, but not the ORDER BY clause. You can, however, sort the final results with an ORDER BY at the end.

Set operations can also be used for special techniques, such as creating sample data, comparing result sets, and combining aggregates.

Coming Up

As we said, we won't presume that you're an expert in all of this. As we introduce the following chapters, we'll also recap some of the basic principles to help you steady your feet.

In the chapters that follow, we'll have a good look at working with the following ideas:

- How to improve the reliability and efficiency of the database tables (Chapter 2)

- How the tables are related to each other and how to work with multiple tables (Chapter 3)

- How to manipulate the values to get more value out of the values (Chapter 4)

- How to summarize and analyze data (Chapter 5)

- How we can save queries and interim results (Chapter 6)

- How to mix data from multiple tables and aggregates (Chapter 7)

- How to work more complex queries by building on top of other queries (Chapters 6, 7, and 9)

- How to add running aggregates and ranking data to our datasets (Chapter 8)

In Chapter 2, we'll make a few changes to the database tables and even add a few more tables to improve its overall design. It won't be perfect, but it will show how a database can be further developed.

## CHAPTER 2

Working with Table Design

SQL databases are, or at least should be, built on some pretty strong principles. Although these principles are sometimes relaxed in real life, the database will be more reliable and more efficient if they're followed as far as possible.

In this chapter, we're going to look at parts of the existing database and how it can be improved using some of those principles. We'll look at

- The basic understanding of normal tables—tables which have been constructed or reconstructed along the core principles

- Modifying tables to follow the principles more closely

- Improving the reliability and integrity of the database by adding additional checks and constraints

- Improving the performance of the database by adding indexes to help find data more efficiently

Of course, we won't be able to make the table perfect: that would take a long time and a lot of experience with the database. You mightn't even be in a position to do this with your database. However, we'll be able to get a better understanding of what makes a database work better.

Understanding Normalized Tables

One of the first things to consider is how individual tables are designed and constructed. The goal is to be sure that each piece of data has a distinct and identifiable place and that the data is in its simplest form.

Mathematicians have a term for anything in its purest form: they say it is **normal**. As with so many mathematical terms, it probably doesn't mean what it looks like: it doesn't mean that it's common, but that it's definitive.

© Mark Simon 2023

M. Simon, *Leveling Up with SQL*, https://doi.org/10.1007/978-1-4842-9685-1_2

Developing a database table often starts with a rough idea of the sort of data to be stored and then goes through a process of **normalization**: examining the features of the table and making changes to fit in better with the requirements of a normal table.

There are various rules and levels of normalization. In general, a normalized database meets the following requirements:

- Data is atomic.

  This means that data is broken down into its smallest practical parts.

- Rows are unordered.

  Of course, you will always see the rows in some sort of order, but the row order is insignificant.

- Rows are unique.

  There shouldn't be any genuine duplicates. Coincidences are, of course, another matter.

- Rows are independent.

  Whatever appears in one row should have no effect on what appears in another row within the table.

- Columns are independent of each other.

  More technically, columns are dependent only on the primary key.

- Columns are of a single type.

  You can't mix data types in a single column. Strictly, you shouldn't be able to mix **domains**, which are sets of acceptable values, but SQL has a hard time checking that.

- Column names are unique.

- Columns are unordered.

  Again, columns will appear in some sort of order—by default, in the order in which the table is designed. However, the column order is insignificant.

One of the design problems that normalization addresses deals with multiple values. If data is to be atomic, and columns are to be independent, how do you manage multiple values? For example, how do you handle sales with multiple sale items or books with multiple genres?

The solution is to put these values into a separate table, one value per row, and let the table refer back to the first table. We will be looking at the relationships between tables, especially in handling multiple values, in the next chapter. The basic idea will be that an additional table will hold multiple values in multiple rows; it will then include a **foreign key**, a reference to a primary key in the first table.

In this chapter, we'll look at some of these principles and see how our database compares against them. When we do find shortcomings, we'll make some changes to the tables themselves and even add a few more tables to make the database more conformant to these ideas. We'll also look at ways of ensuring that the data is more reliable and, to some extent, more efficient.

We'll begin by addressing problems with interdependent columns, which will mean changing some tables. It will also mean adding additional tables. To make these changes less inconvenient, we'll look at creating views to take in the additional tables.

We'll also look at how to work toward a more reliable database by adding additional constraints—data rules—to check what goes into the table in the first place.

Finally, we'll look at adding indexes to improve the performance of the database.

When it comes to the question of multiple values, we'll see more on that in the next chapter, which deals with how tables are related with each other.

No database is perfect, and it won't be our aim to make this one perfect: we'll leave a lot of work undone. It's also quite likely not your job anyway. However, at least we'll get a better understanding of what makes a good database work.

Columns Should Be Independent

One of the fundamental principles of table design is that columns should be independent of each other. That is, changing one column shouldn't necessarily affect another.

However, if you look at customer addresses:

```
SELECT
    id, givenname, familyname,
    street, town, state, postcode
FROM customers;
```

you will see that there is indeed a relationship between some of the address columns.

Id	...	street	town	state	postcode
85	z	1313 Webfoot Walk	Kingston	ACT	2604
355	...	345 Stonecave Road	Kingston	ACT	2604
147	...	Apartment 5A, 129 West 81st Street	Kingston	ACT	2604
112	...	890 Fifth Avenue	Kingston	ACT	2604
489	...	Apartment 42, 2630 Hegal Place	Gordon	ACT	2906
592	...	0001 Cemetery Lane	Rosebery	NSW	1445

~ 303 rows ~

For example, if you change your address from one town to another, you will probably also need to change the postcode and possibly the state. On top of that, people living in the same town probably also have the same postcode; certainly, they will be in the same state.

This creates a maintenance problem:

- Changing address requires changes in *three* columns for a single change.

- It is possible to make a mistake and change only some of the data; this creates an inconsistency, making the data useless.

The correct solution would be to move this data to another table.

Adding the Towns Table

There is an SQL file called towns.sql. This will create and populate a towns table.

The table has the following structure:

```
CREATE TABLE towns (
    id INT PRIMARY KEY, --  auto numbered
    name VARCHAR(...),
    state VARCHAR(...),
    postcode CHAR(4),
    UNIQUE(name,state,postcode)
);
```

28

The table also includes an autonumbering id column, which is the **primary key**. A primary key is a column which uniquely identifies a row. In this case, it's an arbitrary number. The actual details will depend on the DBMS you're using.

Although there will be duplicated names, states, and postcodes, the *combination* will be unique.

The preceding UNIQUE clause also creates an index, which will make searching the table faster. You will learn more about indexes later.

You can run this script now to create and populate the towns table.

Depending on your DBMS, you may need to make sure that you are installing this table into the correct database.

Adding a Foreign Key to the Town

The next step will be to add a column in the customers table to reference the town in the towns table. This is done with an ALTER statement:

```
ALTER TABLE customers
ADD townid INT
    CONSTRAINT fk_customers_town REFERENCES towns(id);
```

Note that the townid column must match the data type of the id column in the towns table, which, in this case, is an integer.

You'll notice that it doesn't actually use the term FOREIGN KEY. It's the keyword REFERENCES that makes it a foreign key: in this case, it references an id in the towns table.

You'll also notice the naming of the foreign key using CONSTRAINT fk_customers_town. Every constraint actually has a name, but you don't have to name it yourself if you're prepared to allow the DBMS to make one up. If so, you can use a shorter form:

```
ALTER TABLE customers
ADD townid INT REFERENCES towns(id);
```

If you already had the column, you could have added the foreign key constraint retroactively with

```
ALTER TABLE customers
ADD CONSTRAINT fk_customers_town FOREIGN KEY(townid)
    REFERENCES towns(id);
```

By default, when you create a new column, it will be filled with NULLs. You could have added a default value instead, but that would be pointless in this case, since everybody lives somewhere else; in some cases, we don't have the customer's address at all.

Update the Customers Table

Now that you have a foreign key column, you will need to fill it with references to the corresponding towns.

If you want to see what these foreign keys should be, you can use a subquery:

```
SELECT
    id, givenname, familyname,
    town, state, postcode,    -- existing data
    (SELECT id FROM towns AS t WHERE    -- new data
        t.name=customers.town
        AND t.postcode=customers.postcode
        AND t.state=customers.state
    ) AS reference
FROM customers;
```

Some of these results will, of course, be NULL, as some of the customers have no recorded address.

id	givenname	familyname	town	state	postcode	reference
85	Corey	Ander	Kingston	ACT	2604	35
355	Joe	Kerr	Kingston	ACT	2604	35
147	Aiden	Abet	Kingston	ACT	2604	35
112	Jerry	Cann	Kingston	ACT	2604	35
489	Justin	Case	Gordon	ACT	2906	135
592	Paddy	Wagon	Rosebery	NSW	1445	386

~ 303 rows ~

A subquery is a query within a query. In this case, it's a simple way of looking up something from another table.

This subquery is a *correlated* subquery: it is run for every row in the main query, using values from the main query to compare to the subquery. That's normally an expensive type of query, but we won't use it very much. It will also be useful for the next step.

You will learn more about subqueries later.

Note that we have aliased the `towns` table in the subquery; that's to make the code easier to read and write. You could also have aliased the `customers` table, but that won't work for all DBMSs in the next step.

We don't just want to look at the reference: we want to copy the reference into the `customers` table. You do that with an UPDATE statement:

```
UPDATE customers
SET townid=(
    SELECT id FROM towns AS t
    WHERE t.name=customers.town
        AND t.postcode=customers.postcode
        AND t.state=customers.state
);
```

The UPDATE statement is used to change values in an existing table. You can set the value to a constant value, a calculated value, or, as in this case, a value from another table.

Here, the same subquery is used to fetch the `id` that will be copied into the `townid` column.

Some DBMSs allow you to alias the `customers` table, which would make the UPDATE statement a little simpler.

A correlated subquery can be expensive, and it's normally preferable to use a join if you can. We could have used a join for the `SELECT` statement, but not all DBMSs cooperate so well with UPDATE statements. Here, the subquery is intuitive and works well, and, since you're only running this once, not too expensive.

Remove the Old Address Columns

Shortly, we will remove the old address columns, but that's going to create an inconvenience later, since the town data is now in a separate table. That means we can't get the customer's details including the full address without joining the tables. To manage this inconvenience, it will be useful to create a view to combine the customer data with the town data.

Create a customerdetails View

A view is a saved query which we can use as a virtual table. In this view, we will simply add all of the columns from the `customers` table except for the town data. For the town data, we will need to join to the `towns` table.

First, we try things out with a simple `SELECT` statement:

```
SELECT
    c.id, c.email, c.familyname, c.givenname,
    c.street,
    --  original values
        c.town, c.state, c.postcode,
    c.townid,
    --  from towns table
        t.name AS town, t.state, t.postcode,
    c.dob, c.phone, c.spam, c.height
FROM customers AS c LEFT JOIN towns AS t ON c.townid=t.id;
```

If you're doing this in Oracle, remember that you can't use AS for the table aliases:

```
SELECT
    ...
FROM customers c LEFT JOIN towns t ON c.townid=t.id;
```

Note that

- We use the `LEFT JOIN` to include customers without an address.

- We alias the `customers` and `towns` tables for convenience.

- The `towns` table has a `name` column, instead of the `town` column. However, in the context of the query, it makes sense to alias it to `town`.

- We've also included the `c.townid` column, which, though it's redundant, might make it easier to maintain.

Once you have checked that the SELECT statement does the job, you can create a view. Of course, you should leave out the old town data, since the whole point is to use the data from the joined data:

```
CREATE VIEW customerdetails AS
SELECT
    c.id, c.email, c.familyname, c.givenname,
    c.street,
    --  leave out the c.town, c.state, c.postcode
    c.townid, t.name AS town, t.state, t.postcode,
    c.dob, c.phone, c.spam, c.height
FROM customers AS c LEFT JOIN towns AS t ON c.townid=t.id;
```

In Microsoft SQL, you need to wrap the CREATE VIEW statement between a pair of GO keywords:

```
--  MSSQL:
GO
    CREATE VIEW customerdetails AS
    SELECT
        c.id, c.email, c.familyname, c.givenname,
        c.street,
        c.townid, t.name as town, t.state, t.postcode,
        c.dob, c.phone, c.spam, c.height
    FROM customers AS c LEFT JOIN towns AS t ON c.townid=t.id;
GO
```

You will learn more about views later.

Drop the Address Columns

To drop the old address columns, you should be able to run the following:

```
--  PostgreSQL, MySQL / MariaDB
    ALTER TABLE customers
    DROP COLUMN town, DROP COLUMN state, DROP COLUMN postcode;

--  Oracle: not DROP COLUMN
    ALTER TABLE customers DROP (town, state, postcode);
```

```
-- SQLite: You need to drop one column at a time
   ALTER TABLE customers DROP COLUMN town;
   ALTER TABLE customers DROP COLUMN state;
   ALTER TABLE customers DROP COLUMN postcode;

-- MSSQL: Doesn't work (yet)
   ALTER TABLE customers
   DROP COLUMN town, state, postcode;
```

Here, we use DROP COLUMN which removes one or more columns and, of course, all of their data, so you would want to be sure that you don't need it anymore. As you've seen earlier, there are some variations in the syntax between DBMSs.

In Microsoft SQL, you will get an error that you can't drop the postcode column because there is an existing constraint. A **constraint** is an additional rule for a valid value.

In this case, there is a constraint called ck_customers_postcode which requires that postcodes comprise four digits only. You won't need that constraint now, especially since you're going to remove the column.

To remove the constraint, run

```
-- MSSQL
   ALTER TABLE customers
   DROP CONSTRAINT ck_customers_postcode;
```

Once you have successfully removed the constraint, you can now remove the columns:

```
ALTER TABLE customers DROP COLUMN town, state, postcode;
```

You will now have removed the extraneous address columns.

Remember, if you drop the wrong column, it is very tricky or impossible to get it back.

Changing the Town

Of course, the whole point of the exercise is you should now be able to move to another town with a single change. We'll try this with customer 42.

First, find the address of customer 42, and, in particular, note the `townid`:

```
SELECT * FROM customerdetails WHERE id=42;
```

You'll get something like this:

id	...	townid	town	State	postcode	...
42	...	846	Kings Park	NSW	2148	...

Note that we're reading from the `customerdetails` view, because the town data is no longer in the `customers` table, though the `townid` is.

Now, change the customer's `townid` to anything you like (as long as it's no more than the highest id in the `towns` table):

```
UPDATE customers SET townid=12345 WHERE id=42;
```

If you now check the same customers:

```
SELECT * FROM customerdetails WHERE id=42;
```

you'll get something like this:

id	...	townid	town	State	postcode	...
42	...	12345	Swan Marsh	VIC	3249	...

If you like, you can set it back to its original value.

Here, we've set the `townid` in the `customers` table, which is where it belongs.

Some DBMSs allow you to take a slightly indirect approach to changing this value and change it via the view:

```
--  Not PostgreSQL or SQLite:
    UPDATE customerdetails SET townid = ... WHERE id=42;
```

As you see, this doesn't include PostgreSQL or SQLite.

Of course, you can't really update a view because it's really just a SELECT statement and doesn't contain any data. Instead, the DBMS tries to work out which table the particular column belongs to and passes the change on to the table. There are times when it can't work that out, such as when you try to update a calculated column. In that case, the update will fail, and you'll have to update the table directly.

Adding the Country

For completeness, you may want to add a reference to a country. You might simply have another column with the country name, but many countries have variations in their names, and you don't want three different versions of the country in the same table.

It's better to have a separate table of countries and include a foreign key to this table. You can add this reference to the customers table, but it might make more sense to add this to the towns table, since it is the town which is located in a country.

You'll note that the countries table has much more than we need for the purpose. You probably don't need things like the country's population or area. However, at some point in the future, you might want the country's currency, time zone, or phone prefix, so it doesn't hurt to have it now. It's not a very big table, and you can just ignore what you don't want, but you can always drop the columns you really don't need.

1. There is an SQL file for another table called countries.sql. It has a number of details, but the two most important details are

    ```
    CREATE TABLE countries (
        id CHAR(2) PRIMARY KEY,
        name VARCHAR(...),
        -- etc
    );
    ```

    Note that the primary key is a two-character string. Every country has a predefined two-character code, generally based on the country's name, either in English or in the country's language. It makes sense to use this as its primary key, rather than making one up. This is an example of a **natural key**: a primary key based on actual data rather than an arbitrary code.

    Run the script to install the table.

2. Add a `countryid` column to the `towns` table, similar to the way you added `townid` to the `customers` table. Remember, the data type must match the preceding primary key:

```
-- PostgreSQL, Oracle, MSSQL, SQLite
   ALTER TABLE towns
   ADD countryid CHAR(2)
   CONSTRAINT fk_town_country REFERENCES countries(id);
```

```
-- MySQL / MariaDB
   ALTER TABLE towns
   ADD countryid CHAR(2) REFERENCES countries(id);
```

3. Update the `towns` table to set the value of `countryid` to `'au'` for Australia or whichever country you choose. This is much simpler than setting it from a subquery:

```
UPDATE towns SET countryid='au';
```

4. You will have to modify your view. First, drop the old version:

```
-- Not Oracle:
   DROP VIEW IF EXISTS customerdetails;
-- Oracle:
   DROP VIEW customerdetails;
```

5. Next, you will have to recreate it with the country name:

```
-- Not Oracle
   CREATE VIEW customerdetails AS
   SELECT
       ...
       c.townid, t.name AS town, t.state, t.postcode,
       n.name AS country
       ...
   FROM
       customers AS c
       LEFT JOIN towns AS t ON c.townid=t.id
```

37

```
            LEFT JOIN countries AS n ON t.countryid=n.id;
--   Oracle
        CREATE VIEW customerdetails AS
        SELECT
            ...
            c.townid, t.name AS town, t.state, t.postcode,
            n.name AS country
            ...
        FROM
            customers c
            LEFT JOIN towns t ON c.townid=t.id
            LEFT JOIN countries n ON t.countryid=n.id;
```

Note

- This includes an additional JOIN to the countries table; to accommodate the longer clause, we have split the JOIN over multiple lines.

- The alias for the countries table has been set to n (for Nation); this is simply because we can't use c as it is already in use.

Additional Comments

You may have noticed that we didn't do anything about the street address column. Strictly speaking, this is also subject to the same issues as the rest of the address, so it would have been better if we did something similar.

However, street addresses are much more complicated, and we don't have so many customers, so we have left them as they are. This leaves us with an imperfect but much improved design.

Improving Database Integrity

So far, we have focused on bringing a table closer to a true normal form by reducing the dependency between columns and repetition of data. This meant adding other tables and foreign keys.

Here, we will explore additional improvements to the integrity of the database. **Database integrity** refers to the quality of the data: Does the data make any sense?

You need to remember that the DBMS really has no idea of what's going on, and it really doesn't care whether you're telling the truth. However, the DBMS is deeply concerned with whether the data is *valid*. That is, whether the data conforms to various rules.

In theory, data belongs to a **domain**—a set of valid values. You should then be able to define a domain for one or more columns. In practice, this feature isn't widely supported in most DBMSs.

On the other hand, you can readily impose **constraints** on a column. A constraint is a data rule.

You'll already know some standard constraint types:

- A data type, such as `INTEGER` or `VARCHAR(16)`, limits the type and range of the data which is acceptable.

- A `NOT NULL` constraint means the value cannot be `NULL`; that is, it's required.

- A `UNIQUE` constraint will disallow a value if another row has the same value already in that column (or combination of columns).

- A `REFERENCES` constraint defines a foreign key; a foreign key *must* match an existing value in another key.

In all cases, of course, there's no guarantee that the value is true—just valid.

If you want to get more specific in your definition of what is valid, there is also the `CHECK` constraint. The `CHECK` is a miscellaneous constraint which allows you to set up your own rules using an expression similar to a `WHERE` clause. Sometimes, these are called **business rules**.

In this section, we'll look at some weaknesses of the database and try to fill in some of the design gaps by adding some constraints.

Much of the following will involve making changes to existing columns. If you're using SQLite, then, sadly, you can't do that. SQLite has very limited ALTER TABLE functionality, and you can't make changes to existing columns. If you really need to make such changes, you would have to go through a more complicated process of dropping a column and creating a new one.

Fixing Issues with a Nullable Column

The saleitems table includes a column called quantity—this is the number of copies you're buying of the book:

```
SELECT * FROM saleitems ORDER BY saleid,id;
```

You'll see something like this:

Id	saleid	bookid	quantity	price
1	1	1403	1	11.5
2	1	1861	1	13.5
3	1	643	[NULL]	18
4	2	187	1	10
5	2	1530	1	12.5
6	2	1412	2	16
~ 13964 rows ~				

However, through an oversight, the column allows NULL, which, if you look far enough, you'll find in a number of rows. That doesn't make sense: you can't have a sale item if you don't know how many copies it's for.

It's reasonable to guess that a missing quantity suggests a quantity of 1. You can implement this guess using coalesce():

```
SELECT
    id, saleid, bookid,
    coalesce(quantity,1) AS quantity, price
FROM saleitems
ORDER BY saleid, id;
```

Now we'll get the same results, except that the NULLs have been replaced with 1:

id	saleid	bookid	quantity	Price
1	1	1403	1	11.5
2	1	1861	1	13.5
3	1	643	1	18
4	2	187	1	10
5	2	1530	1	12.5
6	2	1412	2	16
~ 13964 rows ~				

As always with the `coalesce()` function, you need to check your assumptions. Is 1 *really* a reasonable guess? In this case, it's unlikely to mean zero copies or any other number, but it all really depends on the situation. For the exercise, we'll just play along...

We certainly don't want to keep doing this every time, so we're going to fix the old values and prevent the NULLs in the future.

What follows won't work with SQLite. However, there is a section after this which is what you might do to make the same changes in SQLite.

Replacing NULL Quantities

First, we'll disallow NULLs. Shortly, we will add a NOT NULL constraint to the `quantity` column. However, we can't do that until we clear out the existing NULLs, because the DBMS will never allow the constraints to be violated, even if the constraints are added later.

Assuming this is OK, we can replace the NULLs with 1:

```
UPDATE saleitems
SET quantity=1
WHERE quantity IS NULL;
```

From here, we won't need to use `coalesce()` on existing data, but we need to prevent NULLs in the future.

Setting the NOT NULL Constraint for Quantity

The next thing is to set a `NOT NULL` constraint on the column:

```
--  PostgreSQL
    ALTER TABLE saleitems ALTER COLUMN quantity SET NOT NULL;
--  MySQL/MariaDB
    ALTER TABLE saleitems MODIFY quantity INT NOT NULL;
--  MSSQL
    ALTER TABLE saleitems ALTER COLUMN quantity INT NOT NULL;
--  Oracle
    ALTER TABLE saleitems MODIFY quantity NOT NULL;
--  Not Possible in SQLite
```

Earlier, the `ALTER TABLE` statement was used to add or remove a column. You can also use it to make changes to an existing column. Here, we use it to add a `NOT NULL` constraint.

As you've seen earlier, each DBMS has its own subtle variation on the `ALTER TABLE` statement.

Setting a DEFAULT for Quantity

In principle, whatever caused the NULLs to appear may happen again, only now it will generate an error. Better still, we should supply a default of 1 in case the quantity is missing in a future transaction:

```
--  PostgreSQL
    ALTER TABLE saleitems
    ALTER COLUMN quantity SET DEFAULT 1;
--  MySQL/MariaDB
    ALTER TABLE saleitems
    MODIFY quantity INT DEFAULT 1;
--  MSSQL
    ALTER TABLE saleitems
    ADD DEFAULT 1 FOR quantity;
```

```
-- Oracle
   ALTER TABLE saleitems
   MODIFY quantity DEFAULT 1;
-- Not Possible in SQLite
```

The DEFAULT value is the value used if you don't supply a value of your own. The column doesn't have to be NOT NULL, and NOT NULL columns don't have to have a DEFAULT. However, in this case, it's a reasonable combination.

Again, note that each DBMS has its own subtle variation on the syntax.

Adding a Positive CHECK Constraint for Quantity

While we're fine-tuning the behavior of the quantity column, you can protect against another possible error. In principle, integers can include negative values, which would make no sense at all for quantity. Even zero wouldn't be appropriate in this case.

You can protect against an out-of-range value using a CHECK constraint:

```
CHECK (quantity>0)
```

You could also impose an upper limit by using the BETWEEN expression:

```
CHECK (quantity BETWEEN 1 AND 5)
```

Remember that BETWEEN is inclusive.

However, you have to be careful in imposing arbitrary limits, such as a maximum of 5, because, say, 6 is not out of the question. We'll leave that decision to some future time.

To add the CHECK constraint, again, you use ALTER TABLE:

```
-- PostgreSQL
   ALTER TABLE saleitems
   ADD CHECK (quantity>0);
-- MySQL/MariaDB
   ALTER TABLE saleitems
   MODIFY quantity INT CHECK(quantity>0);
-- MSSQL
   ALTER TABLE saleitems
   ADD CHECK(quantity>0);
-- Oracle
   ALTER TABLE saleitems
```

43

```
    MODIFY quantity CHECK(quantity>0);
-- Not Possible in SQLite
```

We've now made the quantity column more reliable.

Combining the Changes

In some DBMSs, it is possible to combine the changes in a single ALTER TABLE statement:

```
-- PostgreSQL
    ALTER TABLE saleitems
        ALTER COLUMN quantity SET NOT NULL,
        ALTER COLUMN quantity SET DEFAULT 1,
        ADD CHECK (quantity>0);
-- MySQL/MariaDB
    ALTER TABLE saleitems MODIFY quantity INT
        NOT NULL
        DEFAULT 1
        CHECK(quantity>0);
-- Oracle
    ALTER TABLE saleitems MODIFY quantity
        DEFAULT 1
        NOT NULL
        CHECK(quantity>0);
-- Not Possible in MSSQL
-- Not Possible in SQLite
```

Since you don't actually make this sort of change terribly often, you lose nothing if you keep the steps separate.

Making the Changes in SQLite

As you see, SQLite has very limited ability to make changes. Generally, SQLite can only make the following changes to a table:

- Add a column

- Rename a column

- Drop a column

However, that's enough to make the changes we want, as long as we're happy with a different column order.

To make all of the preceding changes

1. Rename the original quantity column:

    ```
    ALTER TABLE saleitems
    RENAME quantity TO oldquantity;
    ```

2. Add a new quantity column with the required properties:

    ```
    ALTER TABLE saleitems
    ADD quantity INT NOT NULL DEFAULT 1 CHECK(quantity>0);
    ```

3. Copy the data from the old column to the new one:

    ```
    UPDATE saleitems
    SET quantity=oldquantity;
    ```

4. Drop the old column:

    ```
    ALTER TABLE saleitems
    DROP oldquantity;
    ```

The new column will be at the end, which is not where the original was, but that's not really a problem.

Other Adjustments

As often in the development process, it's not hard to get something working, but the main effort goes into making it working just right. Here are some suggestions to improve both the integrity and the performance of the database.

We'll talk about indexes in the next section: they help in making the data easier to search or sort.

table	column	suggestion
Customers		
	height	CHECK (height>0) – or height BETWEEN 60 and 260
	dob	CHECK (dob<current_timestamp)
	registered	CHECK (registered<current_timestamp)
Authors		
	names	INDEX
	dates	CHECK (born<died)
	gender	CHECK (gender IN('m','f'))
		CHECK (givenname IS NOT NULL OR familyname IS NOT NULL)
Books		
	authorid	INDEX
	title	INDEX
	published	CHECK (published < year(current_timestamp))
	price	CHECK (price>=0)
Sales		
	total	CHECK (total>=0)
	ordered	CHECK (ordered<current_timestamp)
	customerid	INDEX
		CHECK (shipped>=ordered)
Saleitems		
	saleid	INDEX
	bookid	INDEX
	quantity	NOT NULL CHECK(quantity>0) DEFAULT 1
	price	CHECK(price>=0)

You'll notice that some of the CHECK constraints aren't associated with a single column. Some constraints are more concerned with how one column relates to another column.

We certainly won't address all of these suggestions here. After all, this isn't a real working database, and it's quite possibly not your job anyway. We'll just look at two more.

Ensuring the Prices Are Not Negative

If it were possible to define a data type as nonnegative, we should use that type for many columns. For example, MySQL/MariaDB has a data type called UNSIGNED INT: being unsigned it will always be zero or positive, which is handy for some counters, as well as for quantities.

The alternative, of course, is to use a CHECK constraint to restrict the values. We'll build that into the price of books.

One thing you might consider is a suitable minimum or maximum price, for which you might use a BETWEEN condition. However, that might change over time, so it's not always practical.

What we'll do here is just make sure that the price is never less than zero. We will allow zero in case we have something we're prepared to give away, but the price should never be less than zero.

Adding that constraint is simple, but again the syntax varies between DBMSs:

```
-- PostgreSQL
   ALTER TABLE books ADD CHECK (price>=0);

-- MySQL/MariaDB
   ALTER TABLE books MODIFY price INT CHECK(price>=0);

-- MSSQL
   ALTER TABLE books ADD CHECK(price>=0);

-- Oracle
   ALTER TABLE books MODIFY price CHECK(price>=0);
```

Again, to do this with SQLite, you can follow the steps for the quantity in saleitems earlier.

Ensuring That an Author Is Born Before Dying

Whenever you have multiple columns with similar data, you run the risk of having them the wrong way round.

One notorious example is the name: you sort the name as `familyname`, then `givenname`, so it makes sense to store it that way, but you often view it the other way. It's not hard to imagine how the name might be entered the wrong way.

The `authors` table has two dates: the `born` date and the `died` date. In this case, they're stored chronologically, but it would still be good to make sure.

Here, we'll add a **table constraint**—a constraint which is applied to the table rather than to an individual column. Column constraints can also be added this way, but some constraints don't apply to single columns.

If we had the chance to recreate the table from scratch, we'd do something like this:

```
CREATE TABLE authors (
    id int PRIMARY KEY,      -- auto-numbered as per DBMS
    -- givenname, othernames, familyname,
    born DATE,
    died DATE,
    -- gender, home,
-- Table Constraint:
    CONSTRAINT ck_author_dates CHECK(born<died)
);
```

Here, the constraint appears as an additional property, typically, though not necessarily, at the end.

Given that the table already exists and that it already has data, that opportunity has passed us by. However, we can add the table constraint retroactively.

First, of course, you need to check for any data which might violate the constraint:

```
SELECT * FROM authors WHERE born<died;
```

There shouldn't be any. If there are, then you're on your own. You'll have to do your own research on what the correct dates should be, or, if you're desperate, you can set them to `NULL`.

The next step would be to add the table constraint:

```
ALTER TABLE authors ADD CHECK (born<died);
```

Unlike adding a column constraint, the various DBMSs all use the same syntax—except, of course, for SQLite. There is no simple method for adding a table constraint in SQLite. Complex methods include dropping and recreating the whole table similar to dropping and recreating a column or tampering with the internals of the database, which is definitely not for the fainthearted.

Adding Indexes

SQL doesn't define what order a table should be in. That leaves it up to the DBMS to store the table in any way it deems most efficient.

The problem is that when searching for a particular row, it could be anywhere, and the only way to find it is to look through the whole table and hope that it doesn't take too long.

If, on the other hand, the table were in order, it would be much easier to find what you're looking for. However, even if it's in order, it's just as likely to be in the order of the wrong thing.

For example, even if the `customers` table is in, say, `id` order, it doesn't help when searching by `familyname`. If it's in `familyname` order, it doesn't help when searching by `phone`.

The solution is to leave the table alone and then supplement the table with one or more **indexes**. An index is an additional listing which is in search order, together with a pointer to the matching row in the table.

For example, the `customers` table has an index for the `familyname`. When the time comes to search on the `familyname`, the DBMS automatically looks up the index instead, finds what it wants, and goes back to the real table to fetch the rest of the data.

There are two costs to having an index:

- Each index takes up a little more space in the database.

- Every time you add or change a row in the table, each index will also need to be updated.

For this reason, you will only find an index on a column if it has been specifically requested in the table design. And you would only include an index if you considered the improvement in search ability to be worth the cost in storage and management.

There are two additional indexes which are automatically included:

- Any UNIQUE column is always indexed; the best way to prevent duplicated values is to keep an ordered list of existing values.

- The primary key is always indexed; by definition, it is a unique identifier, which you would presumably search often.

Another type of column which might be worth considering is a foreign key. That's because it will, of course, be heavily involved in searching and sorting.

There is some discussion in learned circles as to the merits of indexing foreign keys. Overall, it appears to be a good idea, and you would probably do well consider adding an index to each of them.

Any other column would be a matter of judgment. At least it's not hard to change your mind about adding or removing an index at some point in the future.

Some DBMSs do include the ability to store the table in order of one column or the other. This is called a **clustered index or an index organized table**. In some DBMSs, such as Microsoft SQL, the clustering is permanent (the DBMS ensures that the table is maintained in that order); in some others, it is temporary (the DBMS sorts the table once, but you'll have to do it again in the future).

Here, we're ignoring clustering. In any case, you still can't keep the table in multiple orders, so you'll need indexes anyway.

Adding an Index to the Books and Authors Tables

One column which you might search routinely is the title column in the books table. To add an index, use CREATE INDEX:

```
CREATE INDEX ix_books_title
ON books(title);
```

The ON clause identifies the table and the columns you want listed.

It is possible to index multiple columns in a single statement, but that doesn't create multiple separate indexes. Instead, you create an index on the combined value. For example:

```
CREATE INDEX ix_authors_name
ON authors(familyname, givenname, othernames);
```

This will create a single index of the authors' familyname, givenname, and othernames.

Even though the index is built around all three parts of the author's name, it will still be used if you just search for, say, the familyname. However, using a partial index that way presumes that you're at least using the first components of the index, which is why the columns are in that order.

Note that in both statements earlier, the index has been given a name. There are no rules for what that name should be, but developers have their own patterns. For example, the preceding pattern is something like

```
ix_table_columns
```

This isn't a rigid rule, but it makes things easier to work with.

Why does the index need a name anyway? Most of the time, you don't really care. However, there are two reasons:

- Everything stored in the database, including maintenance objects, must have a unique name for internal management.

- If you ever need to drop an index, you need to use its name to identify it.

Even if you succeed in creating an anonymous index, the DBMS will automatically assign its own name, which isn't always a very pretty name.

Another index you might consider is on the foreign key authorid in the books table. You can add it with

```
CREATE INDEX ix_books_authors
ON books(authorid);
```

Of course, you might also include an index on customer details or other details.

Creating a Unique Index

There are some columns where you might not expect duplicated values. For example, in your customer table, you wouldn't expect two customers to have the same email address, especially if they're expected to log in using the email address. Similarly, you mightn't expect different customers to have the same phone numbers.

Other columns, such as the family name or date of birth, should happily allow for duplicates: duplicated values are simply a coincidence.

The customers table has already protected against duplicated email addresses by including the UNIQUE property on that column. We will do the same with the phone number.

Before you do, however, you will need to make sure that there aren't any existing duplicates. SQL will refuse to do anything which violates the constraints of the database, so if you have duplicated phone numbers, you won't be able to add a UNIQUE constraint until those duplicates are resolved.

To find duplicates, you use an aggregate group query. For example, you might want to look for customers with duplicate names:

```
SELECT familyname, givenname, count(*) AS number
FROM customers
GROUP BY familyname, givenname;
```

By grouping the names, you can count how many times they appear. Of course, since you're only interested in those that appear more than once, you can filter the results with a HAVING clause:

```
SELECT familyname, givenname, count(*) AS number
FROM customers
GROUP BY familyname, givenname
HAVING count(*)>1;
```

You should see a short list of candidates:

familyname	Givenname	number
Free	Judy	2
Mate	Annie	2
Christmas	Mary	2
Tuckey	Ken	2
Ander	Corey	2
Dunnit	Ida	2
Bearer	Paul	2
Bell	Terry	2

In the preceding query, the rows are grouped by familyname and givenname and summarized. The HAVING clause filters for those groups where there are more than one instance. The SELECT clause then outputs those names and the number of instances.

You don't need the count(*) in the SELECT clause, of course, but it helps to make the result clearer.

Of course, it's no problem if you find duplicate family names: many people have the same name as someone else. However, it can be if you find duplicate phone numbers:

```
SELECT phone, count(*) AS number
FROM customers
GROUP BY phone
HAVING count(*)>1;
```

phone	number
[NULL]	17

In this case, there are no duplicates. What appear to be duplicates are NULLs, because there are multiple NULLs in the table. They don't count.

If you do find duplicates, then you have your work cut out for you in trying to work out whether these duplicates are legitimate. You might even conclude that duplicate phone numbers are OK, so you wouldn't go ahead with the next step.

Assuming that duplicates are *not* OK, to protect against duplicates, you add a UNIQUE INDEX:

```
--  Not MSSQL
    CREATE UNIQUE INDEX uq_customers_phone
        ON customers(phone);
```

Microsoft SQL has a quirk which regards multiple NULLs as duplicates,[1] so you will need this workaround:

```
--  MSSQL
    CREATE UNIQUE INDEX uq_customers_phone
        ON customers(phone)
    WHERE phone IS NOT NULL;
```

Note that this time the index name begins with uq as a reminder that this is a unique index. Again, there are no rules for how to name the index, but this one follows a common and understandable pattern.

Whether or not you really want to disallow duplicate phone numbers is another question. Two customers from the same household or organization may well share the same phone number, so disallowing them would be problematic. This is an exercise in *how* to disallow duplicates, but not necessarily on *whether* to disallow duplicates. That's something best left to the needs of the individual database.

Review

A well-designed SQL database needs to follow a few rules to ensure that the data can be relied upon. There is no guarantee that the data is *true*, but the data will at least be *valid*.

Normal Form

A table which follows certain design principles is said to be in a normal form. This doesn't mean that it's commonplace, but rather that it is in a definitive form.

---

[1] This is odd, since constraints normally ignore NULLs, and NULL doesn't match NULL anyway.

Normalized tables include the following properties:

- Data is atomic.

- Rows are unordered.

- Rows are unique.

- Rows are independent of each other.

- Columns are independent of each other.

- Columns are of a single type.

- Column names are unique.

- Columns are unordered.

Multiple Values

One issue in developing tables is how to handle multiple values and recurring values. In general, the solution is to have additional tables and to link them using foreign keys.

Altering Tables

When restructuring or hardening a database, you need to make changes to existing tables and columns. The ALTER TABLE statement can be used to

- Add extra columns, including foreign keys

- Drop columns

- Add or drop constraints

- Add or drop indexes

Constraints include adding NOT NULL, defaults, and additional CHECK constraints.

Views

A view is a saved SELECT statement. One reason to create a view is for the convenience of having data from one or more tables in one place.

Sometimes, when you create a view with combined data, you end up with a result which no longer follows all the rules of normalization. In the trade, this would be referred to as **denormalization**.

Denormalized data is generally a bad way to maintain data, but very often a convenient way to extract data. In this sense, it is the best of both worlds: the original data is still intact in the original tables.

Some DBMSs include the ability to update data in view. In fact, the update doesn't affect the view at all, but is rather passed on to the underlying tables.

Indexes

An index is a supplement to a table which stores the selected data in order, together with a reference to the data in the original table. Using the index, the DBMS can search for data more quickly.

Indexes are automatically created for primary keys and unique columns. You can add an index on any other column.

Indexes have some costs, so they shouldn't be added for no reason. Costs include storage and maintenance.

Unique indexes can be added to ensure that values in a particular column, or combination of columns, are unique.

The Final Product

After you've made the changes to the table structures, your database will resemble the design in Figure 2-1.

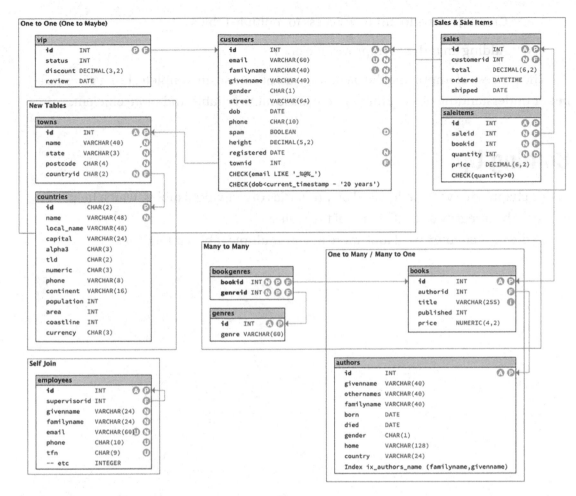

Figure 2-1. *The Improved Database Design*

Summary

In this chapter, we focused on the properties of individual tables and looked for ways to make the database more reliable and more efficient.

We looked at

- The principles of normalized SQL tables

- How multiple values are handled in normalized tables

- Altering tables to improve their reliability and to better fit the principles of normal tables

- Creating views to improve access to multiple tables

- Adding an index to improve efficiency

The process of improving the database was, of course, incomplete, but it gives us a better understanding of what makes a database more reliable and more efficient.

Coming Up

In this chapter, we've been focused on properties of individual tables, which help to improve the integrity and efficiency of the tables.

In the next chapter, we'll look more at how multiple tables interact.

CHAPTER 3

Table Relationships and Joins

A database is not just one table. Well, it can be of course, but any sophisticated database, such as one which you would use to manage an online bookshop, will comprise a number of tables, each handling a different collection of data.

While you can get some useful information from examining individual tables, you will get so much more from combining tables.

In this chapter, we will look at working with multiple tables, how they are related to each other, and how to combine them when the time comes.

Specifically, we'll look at

- What we mean by table relationships and their main types

- How one-to-many relationships are used to manage multiple values

- How one-to-one relationships are used to extend one table with another

- How many-to-many relationships are used to manage more complex multiple values

- How to work with inserting and updating data in multiple tables

We'll look at why the database is structured this way with multiple tables and how we can use joins to combine them into virtual tables.

© Mark Simon 2023
M. Simon, *Leveling Up with SQL*, https://doi.org/10.1007/978-1-4842-9685-1_3

An Overview of Relationships

A well-structured database adheres to a number of important design principles. Two of these are as follows:

- Each table has one type of data only and doesn't include data which rightly belongs in another table.

    For example, the `customers` table doesn't include book details, and the `books` table doesn't include customer details. This would also apply to the `books` and `authors` tables.

    That isn't to say that the `books` table isn't aware of the author at all. We'll look at that in a moment.

- Data is never repeated. The same item of data is not to be found in a different table, nor will it be repeated in the same table.

    For example, if you were to include the author's name and other details in the `books` table, you would find yourself repeating the same details for other books by the same author.

These two principles are related: if you mix author details with the books, violating the first principle, you will end up repeating the details for multiple books, violating the second principle.

The correct way to manage books and authors is to put author details in a separate table and for the `books` table to include a **reference** to one of the authors. In this way, we say that there is a **relationship** between the two tables.

The same would apply to the `books` and `customers` tables. Since the goal is for customers to buy books, there should be a relationship between these tables as well. However, this relationship is a little more complex, as we shall see later.

There are three main types of relationships:

- A **one-to-many** relationship is between the **primary key** of one table and a foreign key of another.

    For example, there is a one-to-many relationship between `authors` and `books`: one author can have many books, and many books can have the same author.

- A **one-to-one** relationship is between the **primary key** of one table and the **primary key** of another. Generally, this is rare, and you are more likely to see a variation of this.

  For example, there is a `vip` table of additional features for customers. For each customer, there can only be one `vip` entry, and there is a (modified) one-to-one relationship between the two tables.

- A **many-to-many** relationship is not a direct relationship, but one that involves a joining table between the two main tables.

  For example, there is a `genres` table which contains possible genre labels for books. Since one book could have many genres, and one genre could apply to many books, there is a many-to-many relationship between the two tables.

  You will see that this is implemented with an additional table.

These relationships might be described as *planned* relationships. They're usually enforced with foreign key constraints, usually involve a primary key, and define a tight structure for the database.

There can also be unplanned relationships. For example, you might consider a relationship between birthdays of customers and authors. That sort of relationship would probably be a coincidence, but might be worth exploring in some situations—maybe Scorpios feel an affinity with other Scorpios.

We'll refer to unplanned relationships as *ad hoc* relationships and look at a few later.

If you have multiple tables in a planned or unplanned relationship, you can examine the combination using a `JOIN`.

One-to-Many Relationship

This is the most common type of relationship between two tables. The relationship is between the primary key of one table and a foreign key in another. However, it's actually implemented as a reference from a foreign key to the primary key.

The relationship is used to indicate a number of possible scenarios. For example:

- One Author has written many Books.

- One Customer has many Sales.

- One Sale contains many Items.

61

Note that the use of the word *many* can imply any number from 0 to ∞.

In the preceding cases, one table is referred to as the **one** table, while the other is referred to as the **many** table, which is not very informative. Sometimes, it is helpful to think of the one table as the **parent** table, while the many table is the **child** table.

A one-to-many relationship is implemented as a reference from the child table to the parent table, for example, for books and authors:

```
CREATE TABLE authors (          -- Parent Table (One)
    id INT PRIMARY KEY
    -- other columns
);
CREATE TABLE books (            -- Child Table (Many)
    id INT PRIMARY KEY,
    bookid INT REFERENCES parent(id)
    -- other columns
);
```

Visually, it looks like Figure 3-1.

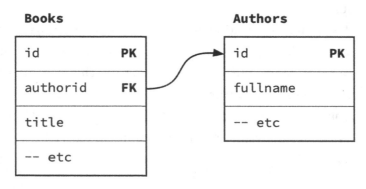

Figure 3-1. *One-to-Many Join*

Note that while the child table has a reference to the parent table, the parent table does *not* have a reference to the child table.

You can combine parent and child tables using a JOIN:

```
-- Not Oracle
    SELECT
        b.id, b.title,        -- etc
        a.givenname, a.familyname    -- etc
```

```
FROM books AS b JOIN authors AS a ON b.authorid=a.id;
```

```
--  Oracle: No AS in table aliases
    SELECT
        b.id, b.title,        -- etc
        a.givenname, a.familyname   -- etc
    FROM books b JOIN authors a ON b.authorid=a.id;
```

This will give you the books with their matching authors:

id	title	givenname	familyname
2078	The Duel	Heinrich	von Kleist
503	Uncle Silas	J.	Le Fanu
2007	North and South	Elizabeth	Gaskell
702	Jane Eyre	Charlotte	Brontë
1530	Robin Hood, The Prince of ...	Alexandre	Dumas
1759	La Curée	Émile	Zola
~ 1172 rows ~			

Note that Oracle has a quirk which disallows using AS for table aliases. If you're using Oracle, you'll need to remember that in the following examples which may include AS.

Remember, if there are anonymous books (books with a NULL for authorid), you will need an outer join:

```
--  Not Oracle
    SELECT
        b.id, b.title,        -- etc
        a.givenname, a.familyname   -- etc
    FROM books AS b LEFT JOIN authors AS a ON b.authorid=a.id;
```

```
--  Oracle: Remember no AS in table aliases
    SELECT
        b.id, b.title,        -- etc
        a.givenname, a.familyname   -- etc
    FROM books b LEFT JOIN authors a ON b.authorid=a.id;
```

This will give you all of the books with or without their authors:

id	Title	givenname	familyname
1868	The Tenant of Wildfell Hall	Anne	Brontë
661	The Narrative of Arthur Gordon Pym ...	Edgar	Poe
91	My Bondage and My Freedom	Frederick	Douglass
848	The Charterhouse of Parma	[NULL]	Stendhal
440	The Princess and the Goblin	George	MacDonald
881	Against Nature	Joris-Karl	Huysmans

~ 1201 rows ~

The one-to-many relationship is the most common relationship between tables.

Remember that SQLite doesn't support RIGHT JOINs. If you want an outer join, you need to put the unmatched row table on the left and make sure to use LEFT JOIN.

Counting One-to-Many Joins

When you're not quite sure how the database is set up, and you're not quite sure whether you've properly joined the right columns to the matching columns, it might be helpful to estimate the number of results you should get from a join.

To get an idea of where we're headed, suppose we have simplified tables of books and authors as in Figure 3-2.

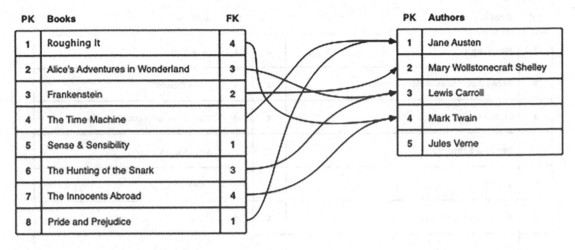

Figure 3-2. Books and Authors

In the previous example, we opted for a LEFT JOIN. When you join a child table to a parent table, you generally have four options:

- Only the matching rows

- Include all of the unmatched children

- Include all of the unmatched parents

- Include all of the unmatched children *and* parents

The first option is, of course, an INNER JOIN, or, more simply, JOIN. The result would look like Figure 3-3.

PK	Books	FK	PK	Authors
1	Roughing It	4	4	Mark Twain
2	Alice's Adventures in Wonderland	3	3	Lewis Carroll
3	Frankenstein	2	2	Mary Wollstonecraft Shelley
5	Sense & Sensibility	1	1	Jane Austen
6	The Hunting of the Snark	3	3	Lewis Carroll
7	The Innocents Abroad	4	4	Mark Twain
8	Pride and Prejudice	1	1	Jane Austen

Figure 3-3. *An Inner Join*

You'll notice that the join doesn't include unmatched books or authors.

The second and third options are LEFT OUTER JOIN or RIGHT OUTER JOIN, depending on whether the unmatched rows are on the left or the right; again, we can simply write LEFT JOIN or RIGHT JOIN. In this case, a LEFT JOIN would include unmatched books, as in Figure 3-4.

PK	Books	FK	PK	Authors
1	Roughing It	4	4	Mark Twain
2	Alice's Adventures in Wonderland	3	3	Lewis Carroll
3	Frankenstein	2	2	Mary Wollstonecraft Shelley
4	The Time Machine			
5	Sense & Sensibility	1	1	Jane Austen
6	The Hunting of the Snark	3	3	Lewis Carroll
7	The Innocents Abroad	4	4	Mark Twain
8	Pride and Prejudice	1	1	Jane Austen

Figure 3-4. *An Outer Join*

Most DBMSs (not including SQLite or MySQL/MariaDB) have a fourth option: include all of the unmatched parents and children. That's called a FULL OUTER JOIN or FULL JOIN to its friends. This would include unmatched rows from both sides as in Figure 3-5.

PK	Books	FK	PK	Authors
1	Roughing It	4	4	Mark Twain
2	Alice's Adventures in Wonderland	3	3	Lewis Carroll
3	Frankenstein	2	2	Mary Wollstonecraft Shelley
4	The Time Machine			
5	Sense & Sensibility	1	1	Jane Austen
6	The Hunting of the Snark	3	3	Lewis Carroll
7	The Innocents Abroad	4	4	Mark Twain
8	Pride and Prejudice	1	1	Jane Austen
			5	Jules Verne

Figure 3-5. *A Full Join*

In this case, we went for the LEFT JOIN because the child table was on the left, and we wanted all of them with or without matches.

Despite the apparent symmetry, all joins are not equal. When you join a child to a parent table, the number of results will generally reflect the child table. That's because many of the children would share the same parent.

To get a fair estimate of how many results you might expect, therefore, you should start by counting the rows.

To get the number of results in an INNER JOIN, you'll need to count the number of children which match a parent—that is, where the foreign key is NOT NULL:

```
-- Child Inner Join
   SELECT count(*) FROM books WHERE authorid IS NOT NULL;
```

That should get you the number of rows for the INNER JOIN previously:

Count
1172

To count the number of unmatched child rows, you just need to count the ones where the foreign key is NULL:

```
-- Unmatched Children
SELECT count(*) FROM books WHERE authorid IS NULL;
```

That will give you the number of rows missing in the INNER JOIN:

Count
29

If you add this to the number for the INNER JOIN, you'll get the total number of books, which is the number of rows in the child OUTER JOIN earlier.

To get the number of unmatched parent rows is trickier. You'll need to count the number of rows in the parent table whose primary key is *not* one of the foreign keys in the child table:

```
-- Unmatched Parents
SELECT count(*) FROM authors
WHERE id NOT IN(SELECT authorid FROM books WHERE authorid IS NOT NULL);
```

This will give you the number of unmatched parents:

Count
45

The subquery selects for the authors whose id does make an appearance in the books table. The NOT IN expression selects for the others. The reason that the subquery includes the WHERE authorid IS NOT NULL clause is due to a quirk in the behavior of NOT IN with NULLs. This is explained later.

Now, you have all the numbers you need to estimate the number of rows in your join. You can use the following combinations:

JOIN	Calculation
INNER JOIN	INNER JOIN
Child OUTER JOIN	INNER JOIN + Unmatched Children= Children
Parent OUTER JOIN	INNER JOIN + Unmatched Parents
Full OUTER JOIN	INNER JOIN + Unmatched Children + Unmatched Parents

That's the number of rows you can expect from a child outer join: LEFT JOIN or RIGHT JOIN, depending on where you put the child table.

Of course, that's not necessarily the end of it. If you have an inner join, and there are some NULL foreign keys, then you'll end up with fewer than the estimate. If you opt for a parent outer join, then there'll be more rows if you have parents without matching children.

However, this is a good starting point.

The NOT IN Quirk

You'd think that if IN(...) finds the rows which match something, then NOT IN(...) would find the others. That's mostly true, except when the list contains a NULL.

For example, to find the customers in two of the states, you can use

```
SELECT * FROM customerdetails
WHERE state IN ('VIC','QLD');
```

To find customers in the other states, you can use NOT IN:

```
SELECT * FROM customerdetails
WHERE state NOT IN ('VIC','QLD');
```

That's as expected. However, if you include NULL in your list, things get messy. You need to remember how IN(...) is interpreted. For example:

```
SELECT * FROM customerdetails
WHERE state IN ('VIC','QLD',NULL);
```

is equivalent to

```
SELECT * FROM customerdetails
WHERE state='VIC' OR state='QLD' OR state=NULL;
```

That last term `state=NULL` will always fail, since NULL *always* fails a comparison, but that's OK if it matches one of the others.

However, the `NOT IN` version:

```
SELECT * FROM customerdetails
WHERE state NOT IN ('VIC','QLD',NULL);
```

is equivalent to

```
SELECT * FROM customerdetails
WHERE state<>'VIC' AND state<>'QLD' AND state<>NULL;
```

When you negate a logical expression, you not only negate the individual terms, but you also negate the operators between them.

Once again, the term `state<>NULL` always fails, but, since this is now ANDed with the rest, it fails the whole expression.

The moral of this story is that you can't use `NOT IN` if the list contains NULLs.

Creating a Books and Authors View

As a general rule, if you've done something a little complicated, you don't want to do it again. Joining the `books` and `authors` table isn't so complicated, but it's still worth considering a simpler approach.

A SELECT statement can be saved permanently in the database as a **View**. A View is just a saved query. Although you can use it like a table, it doesn't actually save data, so it's cheap.

There is also something called a **materialized view** which does save the results, but it's not meant to be a permanent store of new data. A materialized view can be useful when your view is fairly complex and you have a lot of processing to do with the results; it doesn't have to recalculate the results every time.

To create a view, you use CREATE VIEW:

```
-- Drop old version (Not Oracle)
   DROP VIEW IF EXISTS bookdetails;
-- Drop old version (Oracle)
   DROP VIEW bookdetails;
-- (Re) create view
   CREATE VIEW bookdetails AS
   SELECT
       b.id, b.title, b.published, b.price,
       a.givenname, a.othernames, a.familyname,
       a.born, a.died, a.gender, a.home
   FROM
       books AS b
       LEFT JOIN authors AS a ON b.authorid=a.id
   ;
```

You can drop an existing view using DROP VIEW. For most DBMSs, you can use DROP VIEW IF EXISTS if you're not sure that it exists (yet). Not with Oracle, however.

Microsoft SQL has an additional quirk: CREATE VIEW must be the only statement in its batch, so you need to put the statement between the GO keyword, which marks the end of one batch and the beginning of another:

```
DROP VIEW IF EXISTS bookdetails;
-- MSSQL Only:
GO
   CREATE VIEW bookdetails AS
   SELECT
       b.id, b.title, b.published, b.price,
       b.authorid, a.givenname, a.othernames, a.familyname,
       a.born, a.died, a.gender, a.home
   FROM
       books AS b
       LEFT JOIN authors AS a ON b.authorid=a.id
   ;
GO
```

We've included the `authorid` in case you want to use it to get more author details. Once you have saved a view, you can pretend it's another table:

```
SELECT * FROM bookdetails;
```

You'll get the same results as before with a little less effort.

One-to-One Relationships

A one-to-one relationship associates a single row of one table with a single row of another. It is normally between two primary keys.

If every row in one table is associated with a row in another table, then you can consider the second table as an extension of the first table. If that's the case, why not just put all of the columns in the same table? Reasons include the following:

- You want to add more details, but you don't want to change the original table.

- You want to add more details, but you *can't* change the original table (possibly because of permissions).

- The additional table contains details that may be optional: not all rows in the original table require the additional columns.

- You want to keep some of the details in a separate table so that you can add another layer of security to the additional details.

One-to-Maybe Relationships

Technically, a *true* **one-to-one** relationship requires a reference from both tables to each other. Among other things, it is hard to implement as it might require adding both rows at the same time.

Since a row from table A must reference a row from table B, you would need to have the table B row in place before you add to table A. However, if a row from table B must also reference a row from table A, then you need to add to table A first. That's clearly a contradiction.

One way to do this would be to **defer** the foreign key constraint until after you've added to both tables in either order. Unfortunately, most DBMSs don't let you do this, so you're stuck with this impossible situation.

A more common variation of this relationship is the **one-to-maybe** relationship.[1] This allows the second table to contain data which does not necessarily apply to every row of the first table.

For example, the vip table contains extra details for some of the customers. In SQL, it looks something like this:

```
CREATE TABLE customers (          -- main table
    id INT PRIMARY KEY
    --  main data
);
CREATE TABLE vip (            --  secondary table
    id INT PRIMARY KEY REFERENCES customers(id)
    --  additional columns
);
```

Visually, it looks like Figure 3-6.

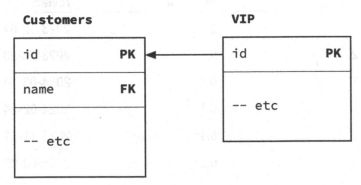

Figure 3-6. *One-to-Maybe Join*

Here, the secondary table contains additional data for *some* of the rows in the main table.

---

[1] One to maybe: My term. Others call it a one-to-zero-or-one, which is less snappy.

Note that this relationship is implemented by making the id in the secondary table both a primary key and a foreign key.

For example, the vip table includes additional features for some customers:

```
SELECT * FROM customers ORDER BY id;
SELECT * FROM vip ORDER BY id;
```

You can see all of the customers, some of whom also have VIP data:

Id	givenname	familyname	...
1	Pierce	Dears	...
2	Arthur	Moore	...
5	Ray	King	...
6	Gene	Poole	...
9	Donna	Worry	...
10	Ned	Duwell	...

~ 303 rows ~

Id	status	discount	review
5	3	0.05	2023-12-01
10	2	0.1	2023-08-23
21	1	0.15	2024-03-14
26	1	0.15	2024-02-24
40	1	0.15	2023-11-11
41	3	0.05	2024-03-09

~ 81 rows ~

You'll notice that there aren't as many rows in the vip table as in the customers table. There might have been, if every customer were a VIP, but not in this case.

You can see how they relate using a join:

```
SELECT c.*, v.*
FROM customers AS c LEFT JOIN vip AS v ON c.id=v.id;
```

This effectively gives you a widened `customers` table:

id	givenname	familyname	...	id	status	discount	review
42	May	Knott	...	42	3	0.05	2024-01-03
459	Rick	Shaw	...				
597	Ike	Andy	...				
186	Pat	Downe	...				
352	Basil	Isk	...	352	1	0.15	2023-08-06
576	Pearl	Divers	...				
~ 303 rows ~							

This gives all of the customers, with either their VIP data or NULLs in the extra columns.

Note

- We need the LEFT JOIN to include non-VIP customers. If you wanted VIP customers only, a simple (inner) JOIN would be better.

- We could have used SELECT *, but using c.*, v.* allows you to decide which tables you are most interested in.

As a special case, you can also select VIP customers only, without additional VIP columns, using

```
SELECT c.*
FROM customers AS c JOIN vip AS v ON c.id=v.id;
```

Here, the inner join selects only VIP customers, and the c.* selects only the customer columns.

Why you would want to do this is, of course, up to you.

Multiple Values

One major question in database design is how to handle multiple values. The principles of properly normalized tables preclude multiple values in a row:

- A single column cannot contain multiple values.

  For example, you shouldn't have multiple phone numbers in a single column, such as 0270101234,0355505678. This would be impossible to sort or search properly and be a real nightmare to maintain.

- You shouldn't have multiple columns with the same role.

  For example, you shouldn't have multiple columns for phone numbers such as phone1, phone2, phone3, etc. It would make searching problematic as you can't be sure which column to search. You would also have the problem of too many or not enough columns.

  Note that you *can* do this if there is a clear distinction in the *type* of phone number. For example, you could legitimately have separate columns for fax (does anybody remember these?), mobile, and landline numbers.

For example, suppose we wish to record multiple genres for a book. Here are two attempted solutions which are *not correct*:

- A column with multiple (delimited) values

  The idea is that the genre column would have multiple genres or genre ids, delimited possibly by a comma. The problem is that the data is not atomic, and this becomes very difficult to sort, search, and update. You will also need extra work to use the data.

- Multiple columns for genres

  You cannot have multiple columns with the same name, so these columns might be called genre1, genre2, etc. Here, the problems are (a) you will either have too many or not enough columns, (b) there is no "correct" column for a particular value, and (c) searching and sorting are impractical.

The problem of recording genres is more complicated, because not only can one book have multiple genres, one genre can apply to multiple books. This is an example of a **many-to-many** relationship.

This cannot be achieved directly between the two tables; rather, it involves an additional table between them.

If you have the courage to look at the script which created and populated the sample database, you'll find a table called booksgenres (not to be confused with the bookgenres table, which is, of course, *completely* different) which does indeed have the genres combined in a single column. This is, of course, cheating.

The booksgenres table is used as a simple method of carrying what will be thousands of rows in two other tables, and there is a rather frightening part of the script later on which pulls this data apart to populate these tables. The booksgenres table will then be dropped to hide the evidence.

This is one case where you might break the rules for the purpose of transferring or backing up the data only. However, the data should *never* stay in this format.

Many-to-Many Relationships

To represent a many-to-many relationship between tables, you will need another table which links the two others.

Such a table is called an **associative** table or a **bridging** table.

It looks like Figure 3-7.

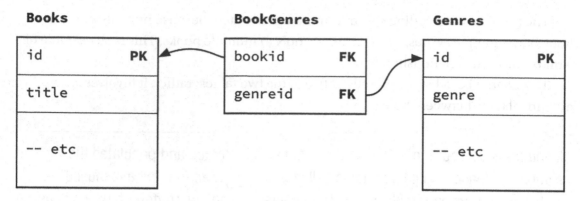

Figure 3-7. *A Many-to-Many Relationship*

Here are two table extracts to illustrate this:

```
-- Book Table
   CREATE TABLE books (
       id int PRIMARY KEY,
       title varchar,
       -- etc
   );
-- Genre Table
   CREATE TABLE genres (
       id int PRIMARY KEY,
       name varchar,
       description varchar
       -- etc
   );
```

The genres table includes a surrogate primary key. It also contains the actual genre name and a description so that the use of the particular genre is clear.

You can see what's in the two tables with simple SELECT statements:

```
SELECT * FROM books;
SELECT * FROM genres;
```

You'll see two apparently unrelated sets of data:

id	authorid	title	published	price
2078	765	The Duel	1811	12.5
503	128	Uncle Silas	1864	17
2007	99	North and South	1854	17.5
702	547	Jane Eyre	1847	17.5
1530	28	Robin Hood, The Prince of Thieves	1862	12.5
1759	17	La Curée	1872	16

~ 1201 rows ~

id	Genre
1	Biology
2	Ancient History
3	Academia
4	Science
5	College
6	Comics

~ 166 rows ~

Neither table refers to the other. Instead, you need an additional table.

The **associative** table will then link books with genres:

```
--  Associative Table
    CREATE TABLE book_genres (
        bookid int REFERENCES books(id),
        genreid int REFERENCES genres(id)
    );
```

You can see what's in this table with the following:

```
SELECT * FROM bookgenres;
```

You'll get a very simple, very boring result:

bookid	Genreid
456	8
789	8
123	52
456	38
789	38
123	80
456	94
356	1
789	113
123	9
1914	1
936	1
1198	1
918	1
456	35
789	68
456	146
789	80
456	101
456	145
1618	2
844	3
~ 8011 rows ~	

This table is a simple table which has one job only: record which books are related to which tables.

Each column must be a foreign key to the other table; otherwise, the whole point of the association is lost. This association allows a book to be associated with multiple genres and a genre to be associated with multiple books.

In the preceding table, for example, book 123 has multiple genres. Book 456 also has two genres. Some of those genres appear for both books and, for all we know, other books later on. That is, one book can have many genres, and one genre can associate with many books.

There is one more requirement. The combination should be unique. There is no point in associating a book with the same genre more than once. Since there is no other data in the table, it would be appropriate to make the combination a compound primary key:

```
-- Associative Table
CREATE TABLE book_genres (
    bookid int REFERENCES books(id),
    genreid int REFERENCES genres(id),
    PRIMARY KEY (bookid,genreid)
);
```

We *might* have designed the table this way:

```
CREATE TABLE bookgenres (
    id INT PRIMARY KEY,
    bookid INT REFERENCES books(id),
    genreid INT REFERENCES genres(id),
    UNIQUE (bookid,genreid)
);
```

which has a UNIQUE constraint on the book/genre combination, as well as a separate primary key. However, by definition, the primary key is already unique, so we can use the more compact design as before.

Note that this is one of the few cases where the primary key is not a single column but a combination of columns. For that reason, the PRIMARY KEY is added separately to the individual columns.

Joining Many-to-Many Tables

To see how books and genres are related, you will need to join these two tables via the associative table.

Normally, you would consider counting the rows in the child table to estimate the number of joined results. In this case, it is the associative table which is the child of both the other tables.

To count the number of rows:

```
SELECT count(*) FROM bookgenres;
```

We already know the result by now:

Count
8011

To join the tables, you only need an INNER JOIN:

```
SELECT *
FROM
    bookgenres AS bg
    JOIN books AS b ON bg.bookid=b.id
    JOIN genres AS g ON bg.genreid=g.id
;
```

This gives a very long list, because the bookgenres table is very long:

bookid	genreid	id	title	...	id	Genre
1732	8	1732	In His Steps	...	8	Fiction
414	8	414	Poesies	...	8	Fiction
241	106	241	Researches in Teutoni…	...	106	Fantasy
247	153	247	The King in Yellow	...	153	Gothic
1914	38	1914	Voyage of the Beagle	...	38	Classics
936	38	936	The Origin of Species	...	38	Classics

~ 8011 rows ~

82

Here, we started the join from the middle, since we're focusing on the associative table. You could just as readily have started on one end:

```
SELECT *
FROM
    books AS b
    JOIN bookgenres AS bg ON b.id=bg.bookid
    JOIN genres AS g ON bg.genreid=g.id
;
```

You'll get the same data, but, since the tables are in a different order, the columns will also be in a different order.

In reality, you end up with too many columns, two of which are duplicated by the join. You can simplify the result as

```
SELECT b.id, b.title, g.genre
FROM
    bookgenres AS bg
    JOIN books AS b ON bg.bookid=b.id
    JOIN genres AS g ON bg.genreid=g.id
;
```

You get a simpler result:

id	title	Genre
1732	In His Steps	Fiction
414	Poesies	Fiction
241	Researches in Teutonic Mythology	Fantasy
247	The King in Yellow	Gothic
1914	Voyage of the Beagle	Classics
936	The Origin of Species	Classics
~ 8011 rows ~		

The book's id is important because it is quite possible for different books to have the same title.

Since you've now listed the specific columns, the column order won't depend on the order of the tables in the join.

By its very nature, the associative table cannot include NULLs for either the bookid or the genreid. As such, there is no need for an outer join.

Summarizing Multiple Values

The resulting table earlier has a very large number of rows, and individual books appear multiple times, associated with different genres. You can combine multiple values using a GROUP BY query.

This could get tricky, because you're trying to summarize not a simple table, but a generated virtual table in the form of a join. You can do this in the form of a Common Table Expression:

```
WITH cte AS (
    SELECT b.id, b.title, g.genre
    FROM bookgenres AS bg
        JOIN books AS b ON bg.bookid=b.id
        JOIN genres AS g ON bg.genreid=g.id
)
--etc
;
```

A Common Table Expression (CTE) saves the results of a SELECT statement into a virtual table, so you can use those results in the next stage. You'll see more on CTEs in Chapters 7 and 9.

You can now summarize the CTE using two functions, count() and string_agg():

```
WITH cte AS (
    SELECT b.id, b.title, g.genre
    FROM
        bookgenres AS bg
        JOIN books AS b ON bg.bookid=b.id
        JOIN genres AS g ON bg.genreid=g.id
)
SELECT
    id, title,
```

```
        count(*) AS ncategories,          -- not really needed
    --  PostgreSQL, MSSQL
        string_agg(genre,', ') AS genres
    --  MySQL / MariaDB
        --  group_concat(genre separator ', ') AS genres
    --  SQLite
        --  group_concat(genre, ', ') AS genres
    --  Oracle
        --  listagg(genre separator ', ') AS genres
FROM cte
GROUP BY id, title
ORDER BY id;
```

You now get a more compact list:

id	title	ncategories	Genres
4	Groundwork of th ...	6	Classics, German Literature, Non ...
5	The Toilers of t ...	7	European Literature, French Lite ...
6	The American Sen ...	12	Classic Literature, European Lit ...
7	Songs of Innocen ...	5	18th Century, Poetry, Classics, ...
9	Behind a Mask, O ...	8	Literature, American, Historical ...
11	Lady Susan ...	4	Romance, Historical Fiction, Cla ...

~ 1200 rows ~

The string_agg(column,separator) function concatenates the values in the column using the separator in between. You'll no doubt be surprised to learn that the DBMSs have variations on this function. As you see, in some DBMSs it is called GROUP_CONCAT or listagg.

Note that we have grouped by both the book's id and title columns. Since the id is a primary key, and therefore unique, there is normally no point in attempting to subgroup as before. However, this is a simple way of including the title for the SELECT clause, since the SELECT clause can only contain grouping columns and summaries.

Combining the Joins

Once we know how to handle books and authors, and books and genres, we can join them all together:

```
--   Not SQLite
SELECT
     b.id, b.title, b.published, b.price,
     g.genre,
     a.givenname, a.othernames, a.familyname,
          a.born, a.died, a.gender, a.home
FROM
     authors AS a
     RIGHT JOIN books AS b ON a.id=b.authorid
     LEFT JOIN bookgenres AS bg ON b.id=bg.bookid
     JOIN genres AS g ON bg.genreid=g.id
;
```

This will give a fairly comprehensive dataset:

id	truncate	...	genre	givenname	othernames	familyname	...
1732	In His Steps ...	...	Fiction	Charles	M.	Sheldon	...
414	Poesies ...	...	Fiction	Stéphane		Mallarmé	...
241	Researches in Te ...	...	Fantasy	Viktor		Rydberg	...
247	The King in Yell ...	...	Gothic	Robert	W.	Chambers	...
1914	Voyage of the Be ...	...	Classics	Charles		Darwin	...
936	The Origin of Sp ...	...	Classics	Charles		Darwin	...

~ 8011 rows ~

In the preceding example, we have included most of the columns from the four tables, omitting the foreign keys and most of the other primary keys.

The tables are joined in a line from the `authors` table to the `genres` table. Since we want all of the books, regardless of whether they have associated authors or genres, we use two outer joins. As it turns out, we see examples of each of the three main join types.

SQLite doesn't support the `RIGHT JOIN`, so this won't work.

You can write the joins starting from the books table if you like:

```
SELECT
    b.id, b.title, b.published, b.price,
    g.genre,
    a.givenname, a.othernames, a.familyname,
    a.born, a.died, a.gender, a.home
FROM
    books AS b
    LEFT JOIN bookgenres AS bg ON b.id=bg.bookid
    JOIN genres AS g ON bg.genreid=g.id
    LEFT JOIN authors AS a ON b.authorid=a.id
;
```

Visually, this appears to put the emphasis on the books table, but it will give exactly the same results as before.

This time, SQLite will be happy.

However, there is an alternative, which takes advantage of the view previously created.

First, you can replace the references to the individual books and authors tables with the bookdetails view:

```
SELECT
    bd.id, bd.title, bd.published, bd.price,
    g.genre,
    bd.givenname, bd.othernames, bd.familyname,
    bd.born, bd.died, bd.gender, bd.home
FROM
    bookdetails AS bd
    LEFT JOIN bookgenres AS bg ON bd.id=bg.bookid
    JOIN genres AS g ON bg.genreid=g.id;
```

This is a simpler query, building on one you've already prepared. Note:

- The bookdetails view is aliased to bd.

- All of the columns come from the bookdetails view except for the genre, which comes from the genres table.

If you want to combine the genre names, you can do that in a CTE and join the results with the view:

```
WITH cte AS (
    SELECT bg.bookid, string_agg(g.genre,', ') AS genres
    FROM bookgenres AS bg JOIN genres AS g ON bg.genreid=g.id
    GROUP BY bg.bookid
)
SELECT *
FROM bookdetails AS b JOIN cte ON b.id=cte.bookid;
```

You now get the complete genre details as well:

id	title	...	genres	givenname	familyname	...
4	Groundwork of ...	...	Classics, German ...	Immanuel	Kant	...
5	The Toilers ...	...	European Literature ...	Victor	Hugo	...
6	The American ...	...	Classic Literature ...	Anthony	Trollope	...
7	Songs of ...	...	18th Century, Poetry ...	William	Blake	...
9	Behind a Mask ...	...	Literature, American ...	A.M.	Barnard	...
11	Lady Susan ...	...	Romance, Historical ...	Jane	Austen	...
~ 1200 rows ~						

You might want to filter the genres. You can do that inside the CTE:

```
WITH cte AS (
    SELECT bg.bookid, string_agg(g.genre,', ') AS genres
    FROM bookgenres AS bg JOIN genres AS g ON bg.genreid=g.id
    WHERE g.genre IN('Fantasy','Science Fiction')
    GROUP BY bg.bookid
)
SELECT *
FROM bookdetails AS b JOIN cte ON b.id=cte.bookid;
```

You'll then get a filtered list:

id	truncate	...	genres	givenname	familyname	...
589	The Story of ...	...	Fantasy ...	Stanley	Waterloo	...
96	Bee: The Pri ...	...	Fantasy ...	Anatole	France	...
880	The Journey ...	...	Fantasy, Science F...	Ludvig	Holberg	...
591	The Story of ...	...	Fantasy ...	E.	Nesbit	...
1938	Histoire Com ...	...	Science Fiction ...	Cyrano	de Bergerac	...
128	The Year 3000...	...	Science Fiction ...	Paolo	Mantegazza	...

~ 163 rows ~

Note that the concatenated genres column has been aliased to genres. That's the same name as the genres table, so you might get confused. The good news is that SQL doesn't, so you can get away with it. On the other hand, if you're worried about that you can always use double quotes: AS "genres". Of course, you can also choose a better alias.

A word of warning, however. When you start using views inside queries, you will have to consider some possible side effects:

- Since the view is not part of the original database, you may lead to some confusion with other users, since views *look* like tables, but aren't with the rest of the tables.

- If you have too many views in a query, the DBMS optimizer may not be able to work out the most efficient plan for running the query. This can be because some views produce more than you need for the next query, and the optimizer may not be able to work out what you really want.

- If there are any changes to the view, they will, of course, affect the outcome of the query.

- These side effects will be more pronounced if you start to create views using other views. It is often safer to create the new view from scratch.

This doesn't mean that you shouldn't use views in your queries—that's the whole point of creating a view. It does mean, however, that you should be careful when piling them up.

Many-to-Many Relationships Happen All the Time

Any time you have one table linked to two others, you have a many-to-many relationship.

For example, the database includes customers, sales, saleitems, and books. All of these tables are joined with foreign key/primary key relationships. That means there is a many-to-many relationship between customers and books: one customer can buy many books, while one book can be bought by multiple customers (not the same one, of course, but a copy of the book).

This means that the sales and saleitems tables, in particular the saleitems table, are taking on the role of associative tables.

There are two main differences between the sales and saleitems tables in this example and the bookgenres table in the previous example:

- The association doesn't need to be unique. It's quite feasible for a customer to buy the same book on another occasion. In fact, it would do the business good if it happened more often.

- The association isn't the whole story. You also want to record other sales details such as the date, the amount paid, and so on.

In this case, the sales tables are not purely associative, since they contain new data of their own, but they are still doing an associative job.

Another Many-to-Many Example

The current database presumes that each book has a single author only. In principle, you could have a book written by multiple authors. That doesn't happen very often in fiction, but is more common in nonfiction.

To do this, you would need to modify your design as follows:

- Remove the authorid column from the books table.

- Create an authorship table which associates books with authors.

There are a few extra tables, not part of the main database, which we can use to make the concept clear:

- `multibooks`: A small table of books without an author id

- `multiauthors`: A table of authors for the preceding books

- `authorship`: An associative table which associates books and authors

The table structure looks like this:

```
CREATE TABLE books (
    id INT PRIMARY KEY,
    title varchar(60)
    -- additional book details NOT including author
);
CREATE TABLE authors (
    id INT PRIMARY KEY,
    givenname varchar(24),
    familyname varchar(24)
    -- additional author details NOT including book
);
CREATE TABLE authorship (
    bookid INT REFERENCES books(id),
    authorid INT REFERENCES authors(id),
    PRIMARY KEY(bookid,authorid)
);
```

An example of a book with multiple authors would be

```
INSERT INTO books(title)
VALUES('Good Omens');
INSERT INTO authors(givenname,familyname)
VALUES
    ('Terry','Pratchett'),
    ('Neil','Gaiman');
```

```
INSERT INTO authorship(bookid,authorid)
VALUES
    (1,1),  --  Good Omens, Terry Pratchett
    (1,2);  --  Good Omens, Neil Gaiman
```

You won't need to run the preceding example, as the data is already in the sample tables.

You can fetch the associated data using something like the following:

```
SELECT *
FROM
    multibooks AS b
    JOIN authorship AS ba ON b.id=ba.bookid
    JOIN multiauthors AS a ON ba.authorid=a.id;
```

You'll get something like this:

id	title	authorid	bookid	id	givenname	familyname
18	The Gilded Age	9	18	9	Charles	Warner
9	The Syndic	7	9	7	Cyril	Kornbluth
21	Seller	7	21	7	Cyril	Kornbluth
8	Wolfbane	7	8	7	Cyril	Kornbluth
12	Takeoff	7	12	7	Cyril	Kornbluth
8	Wolfbane	6	8	6	Frederik	Pohl

~ 31 rows ~

You can also combine the authors for each book using a CTE and an aggregate query:

```
WITH cte AS (
    SELECT
        ba.bookid,
        string_agg(a.givenname||' '||a.familyname,' & ')
            AS authors
    FROM authorship AS ba JOIN multiauthors AS a
        ON ba.authorid=a.id
    GROUP BY ba.bookid
```

```
)
SELECT b.id, b.title, cte.authors
FROM multibooks AS b JOIN cte ON b.id=cte.bookid
ORDER BY b.id;
```

You now have a list of books with all of their authors:

id	title	Authors
1	Man Plus	Frederik Pohl
2	Proxima	Stephen Baxter
3	The Long Mars	Stephen Baxter & Terry Pratchett
4	The Shining	Stephen King
5	The Talisman	Peter Straub & Stephen King
6	The Long Earth	Stephen Baxter & Terry Pratchett
~ 23 rows ~		

The main sample database doesn't include multiple authors simply because it doesn't happen often enough with classic literature to make it worth complicating the sample further.

However, the point is that whenever you have multiple values, you will need additional tables rather than additional columns or compound columns. Multiple values should appear in rows, not columns.

Inserting into Related Tables

As you see, one-to-many tables are very common in an SQL database. This creates additional challenges when you need to add to multiple tables in a single transaction.

These challenges include the following:

- You may need to add to the parent (one) table before you add to the child (many) table.

    This is because the child table will refer to the parent table.

- When adding to the parent table, you need to remember the new primary key.

This becomes a challenge when the primary key is generated by the database itself, so you will need to fetch the new value.

- During the process, one of the steps may fail.

  This may lead to a partially completed operation, adding some invalid data as a residue.

Here, we'll see how that works in practice.

Adding a Book and an Author

The books collection is very light on works in the 20th century and contains nothing by Agatha Christie. Here, we will add one of her books.

This will involve two tables: the books and the authors table as you see in Figure 3-8.

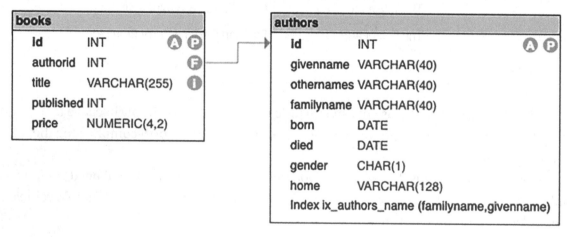

Figure 3-8. Books and Authors

To add a new book to the database

1. Check whether the author already exists in the authors table.

2. If not, add the author to the authors table.

3. If you have just added a new author, fetch its primary key.

4. Else fetch the primary key of the existing author.

5. Using this primary key, add the new book to the books table.

First, we'll check to see whether the author has already been added to the authors table:

```
SELECT * FROM authors WHERE familyname='Christie';
```

It appears not, so we'll need to add her.

Adding an Author

In principle, you would add the new author with the following statement:

```
--  Don't run this yet:
    INSERT INTO authors(givenname, othernames, familyname,
        born, died, gender,home)
    VALUES('Agatha','Mary Clarissa','Christie',
        '1890-09-15','1976-01-12','f',
        'Tourquay, Devon, England');
--  For Oracle, you need to precede the dates with the
--  date keyword: date'1890-09-15', date'1976-01-12'
```

As the comment says, don't run this statement yet. Because the author's id is autogenerated, we'll need to get the new id after inserting the row. You can do a search for it after adding the row, but it may be possible to have the DBMS tell you what the new id is.

Different DBMSs have different methods of getting this id.

For PostgreSQL, you can simply use a RETURNING clause at the end of the INSERT statement:

```
--  PostgreSQL
    INSERT INTO authors(givenname, othernames, familyname,
        born, died, gender,home)
    VALUES('Agatha','Mary Clarissa','Christie',
        '1890-09-15','1976-01-12','f',
        'Tourquay, Devon, England')
    RETURNING id;    --  Take note of this!
```

For MySQL/MariaDB, Microsoft SQL, and SQLite, you run a separate function after the event. Note that you should run *both* statements together by highlighting both before you run:

```
--  MSSQL: Select both statements and run:
    INSERT INTO authors(givenname, othernames, familyname,
        born, died, gender,home)
    VALUES('Agatha','Mary Clarissa','Christie',
        '1890-09-15','1976-01-12','f',
        'Tourquay, Devon, England');
    SELECT scope_identity();         -- Take note of this!

--  MySQL / MariaDB: Select both statements and run:
    INSERT INTO authors(givenname, othernames, familyname,
        born, died, gender,home)
    VALUES('Agatha','Mary Clarissa','Christie',
        '1890-09-15','1976-01-12','f',
        'Tourquay, Devon, England');
    SELECT last_insert_id();         -- Take note of this!

--  SQLite: Select both statements and run:
    INSERT INTO authors(givenname, othernames, familyname,
        born, died, gender,home)
    VALUES('Agatha','Mary Clarissa','Christie',
        '1890-09-15','1976-01-12','f',
        'Tourquay, Devon, England');
    SELECT last_insert_rowid();      -- Take note of this!
```

The additional SELECT statements earlier all fetch the newly generated id.

Oracle, on the other hand, makes it pretty tricky. It does support a RETURNING clause, but only into variables. You can get the newly generated id, but that involves some extra trickery in hunting for sequences. The simplest method really is to select the row you've just inserted using data you've just entered:

```
--  Oracle
    INSERT INTO authors(givenname, othernames, familyname,
        born, died, gender,home)
    VALUES('Agatha','Mary Clarissa','Christie',
```

```
    date '1890-09-15',date '1976-01-12','f',
    'Tourquay, Devon, England');
  SELECT id FROM authors
  WHERE givenname='Agatha' AND othernames='Mary Clarissa'
      AND familyname='Christie';
```

Of course, you don't necessarily need to filter all of the new values: just enough to be sure you've got the right one.

Adding a Book

After that, the rest is easy.

Whether or not you have just added the new author, you can simply search for the authors table to get the author id:

```
SELECT * FROM authors WHERE familyname='Christie';
```

Taking note of the id in particular, you can insert the book with the following statement:

```
-- Use the author's id:
  INSERT INTO books(authorid,title,published,price)
  VALUES ( ... , 'The Mysterious Affair at Styles',
      1920, 16.00);
```

Of course, you will need to supply the correct id in the preceding statement, either from the INSERT statements in the previous section or from the SELECT statement earlier.

Note that we've picked an arbitrary value of 16.00 for the price. It didn't need the decimal part, of course, but it makes the purpose clearer.

Adding a New Sale

Adding a book is simple enough, but tedious. Adding a new sale, however, adds another level of complexity. Remember that a sale comprises a row in the `sales` table and one or more rows in the `saleitems` table, as you see in Figure 3-9.

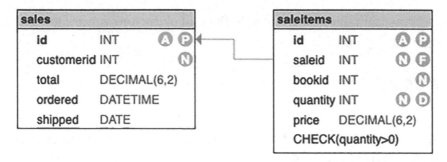

Figure 3-9. *Sales and Sale Items*

The process would be

1. Create a new row in the `sales` table.

2. Fetch the primary key of the new sale.

3. Add one or more rows in the `saleitems` table, using the previous primary key to refer to the new sale. This will include the `bookid` and `quantity`, but the `price` will need to be fetched separately.

4. For each sale item, fetch the price of each item from the `books` table, using the `bookid`.

5. Update the new sale with the total value of the new sale items.

Here, we will work with adding a new sale.

For our sample data, we'll use the following values:

Data	Value
Customer ID	42
Book IDs	123, 456, 789
Quantities	3, 2, 1

We'll also use `current_timestamp` for the date.

Adding a New Sale in the Sales Table

Adding the main sale is the easy part, but, again, we'll need the new id. To add a sale, we can use

```
-- PostgreSQL
   INSERT INTO sales(customerid, ordered)
   VALUES (42,current_timestamp)
   RETURNING id;

-- MSSQL
   INSERT INTO sales(customerid, ordered)
   VALUES (42,current_timestamp);
   SELECT scope_identity();

-- MySQL / MariaDB
   INSERT INTO sales(customerid, ordered)
   VALUES (42,current_timestamp);
   SELECT last_insert_id();

-- SQLite
   INSERT INTO sales(customerid, ordered)
   VALUES (42,current_timestamp);
   SELECT last_insert_rowid();

-- Oracle
   INSERT INTO sales(customerid, ordered, total)
   VALUES (42,current_timestamp,0);
   SELECT id FROM sales WHERE id=42 AND total=0;
```

For Oracle, we've taken a slightly different approach in including a dummy total of 0. When the sale is fully added, the value shouldn't be zero, so we're using it as a temporary placeholder to help identify the new sale.

Remember that we'll need to remember the new sale id.

Adding the Sale Items and Getting the Prices

Armed with the new sale id, the rest is simple. To add the sale items, we can use

```
-- Not Oracle
   INSERT INTO saleitems(saleid, bookid, quantity)
   VALUES
         ( ... , 123, 3),
         ( ... , 456, 1),
         ( ... , 789, 2);

-- Oracle
   INSERT INTO saleitems(saleid, bookid, quantity)
   VALUES ( ... , 123, 3);
   INSERT INTO saleitems(saleid, bookid, quantity)
   VALUES ( ... , 456, 1);
   INSERT INTO saleitems(saleid, bookid, quantity)
   VALUES ( ... , 789, 2);
```

Remember to use your new sale id in the preceding statements.

Also, remember that Oracle doesn't like multiple values in a single INSERT statement, which is why there are multiple statements. You can use that for the other DBMSs if you prefer, but it's not necessary.

The prices come from another table. You can fetch those prices into the new sale items using a subquery:

```
UPDATE saleitems
SET price=(SELECT price FROM books WHERE
    books.id=saleitems.bookid)
WHERE saleid = ... ;
```

The correlated subquery fetches the price from the saleitems table for the matching book (WHERE books.id=saleitems.bookid).

The WHERE clause in the main query ensures that only the new sale items get the prices. This is important because you don't want to copy the prices into the old sale items: there might have been a price change since the older sales were completed, and you shouldn't let that change affect old transactions.

Completing the Sale

Finally, you should find the total for the sale items and put that into the new sale.

To get the total for the sale items, you can use an aggregate query:

```
SELECT sum(quantity*price)
FROM saleitems
WHERE saleid = ... ;
```

The result would be correct, but it would also be incomplete. What's missing is the tax and any VIP discount applicable.

Let's assume a tax of 10%—it varies from country to country, of course, so you might want to make an adjustment. That means you'll end up paying (1 + 10%) times the total:

```
SELECT sum(quantity*price) * (1 + 0.10)
FROM saleitems
WHERE saleid = ... ;
```

In real life, of course, you would simply write 1.1, but the preceding expression is a reminder of where the value came from and how you might adapt it for different tax rates.

The VIP discount depends on the customer. You can read that from the VIP table:

```
SELECT 1 - discount FROM vip WHERE id = 42 ;
```

The reason you subtract it from 1 is that it's a discount: it comes off the full price.

You can use that in a subquery with the calculated total:

```
SELECT
    sum(quantity*price)
    * (1 + 0.1)
    * (SELECT  1 - discount FROM vip WHERE id = 42)
FROM saleitems
WHERE saleid = ... ;
```

except not necessarily. Some customers aren't VIPs, so the subquery might return a NULL. That would destroy the whole calculation. Since a missing VIP value means no discount, we should coalesce the subquery to 1:

```
SELECT
    sum(quantity*price)
    * (1 + 0.1)
    * coalesce((SELECT  1 - discount FROM vip
        WHERE id = 42),1)
FROM saleitems
WHERE saleid = ... ;
```

Finally, we can put this value into the sales table:

```
UPDATE sales
SET total = (
    SELECT
        sum(quantity*price)
        * (1 + 0.1)
        * coalesce((SELECT  1 - discount FROM vip
            WHERE id = 42),1)
    FROM saleitems
    WHERE saleid = ...
)
WHERE id = ... ;
```

There's a lot going on here. First, the UPDATE query sets a value to a subquery, which, in turn, uses a subquery to fetch a value. You'll also find that the query uses the sale id twice, once to filter the sale items and once to select the sale.

Review

A main feature of SQL databases is that there are multiple tables and that these tables are related to each other.

Relationships are generally established through primary keys and foreign keys which reference the primary keys in related tables. The foreign key is normally in the form of a constraint, which guarantees that the foreign key references a valid primary key value in the other table, if not necessarily the correct one.

There may also be *ad hoc* relationships which are not planned or enforced.

Types of Relationships

There are three main relationship types:

- **One-to-many** relationships are between a foreign key in one table, often called a **child** table, and a primary key in another, the **parent**. Generally, one parent can have many children.

  This is the most common type of relationship.

- **One-to-one** relationships are between primary keys in both tables. The primary key in one table doubles up as a foreign key in the other.

  A true one-to-one relationship requires both primary keys to be foreign keys to the other table. In practice, this is difficult to implement, and the foreign key is normally on one table only. This may informally be called a **one-to-maybe** relationship.

- **Many-to-many** relationships allow a row in one table to relate to many rows in the other, as well as the other way around. Since columns can only have single values, this relationship is created through another table, often called an **associative** table, with a pair of one-to-one relationships.

  In any reasonably sized database, the fact that there are many tables in one-to-many relationships results in many-to-many relationships.

It's a basic principle in a database that a column shouldn't have multiple values and that you shouldn't have multiple columns doing the same job. The way to handle multiple values is with additional tables, either in one-to-many or many-to-many relationships.

Joining Tables

When there is an established relationship between tables, you can combine their contents using joins.

Sometimes, you may want to count the number of expected results to check whether your join type matches what you want.

When you do join tables, you often end up with several rows with the same repeated data coming from the parent table. You may be able to simplify this by grouping on parent data and aggregating on the child data. Because you can only select what you summarise, you may need to join the results again to get more details.

Views

Selecting what you want from multiple related tables can be inconvenient. You can save your complex joined query in a view for future use and use it as you might a simple table afterward.

Inserting into Related Tables

Often, inserting into a table is not so simple. In some cases, if it's a child table, you may need to also insert into the parent table so that the child can reference it.

In other cases, such as when the parent table is used as a container for child data, you may need to insert into multiple tables in a number of steps.

This process can be complicated if the primary key is autogenerated.

Summary

In this chapter, we looked at how multiple tables are related through foreign keys matching with primary keys. We also looked at different types of relationships and why tables were designed this way.

Using this, we were able to combine tables using one or more joins to match rows from one table to another. We looked at different types of joins and when you might choose between them.

Coming Up

Most of the data we've worked with have been simple values, though in a few cases we calculated values such as tax and discounts.

In the next chapter, we're going to take a small detour and concentrate on performing calculations in SQL.

CHAPTER 4

Working with Calculated Data

No doubt, you will have seen calculations before now. SQL allows you to include calculated data in your queries.

In this chapter, we'll look at some important ideas about the different types of data and how they can be calculated.

Don't get too carried away with calculations in your SQL. The database is more concerned with maintaining and accessing raw data. However, it's useful to be able to take your raw data and make it more useful to the task at hand.

DBMSs vary widely in their ability to perform calculations. This is especially the case with functions, which vary not only in scope but even in what the DBMSs call them.

In particular, SQLite has a very limited ability to perform calculations, particularly with functions.

In this chapter, we'll be working with various types of data, including strings. If you're using MariaDB/MySQL, we thoroughly recommend that you set your session to ANSI mode, so that string behavior works as for standard SQL.

You can begin your session with SET SESSION sql_mode = 'ANSI';

© Mark Simon 2023
M. Simon, *Leveling Up with SQL*, https://doi.org/10.1007/978-1-4842-9685-1_4

Calculation Basics

We'll look at more details later, but here is an overview of how calculations work in SQL.

You can calculate values based on individual columns or multiple columns. For example:

```
SELECT
    height/2.54,                -- single column
    givenname||' '||familyname        --  multiple columns
    --  givenname+' '+familyname          --  MSSQL
FROM customers;
```

This gives you

?column?	?column?
66.339	May Knott
67.283	Rick Shaw
60.236	Ike Andy
69.291	Pat Downe
61.575	Basil Isk
69.409	Pearl Divers
~ 303 rows ~	

You can also "hard-code" values or get them from subqueries:

```
SELECT
    'active',                  -- hard-coded
    (SELECT name FROM towns WHERE id=townid)  --  sub query
FROM customers;
```

You get

?column?	?column?
active	Kings Park
active	Richmond
active	Hillcrest
active	Guildford
active	Wallaroo
active	Broadwater

~ 303 rows ~

There are also some built-in functions:

```
SELECT
    upper(familyname)         --  upper case function
FROM customers;
```

This gets you

?column?
KNOTT
SHAW
ANDY
DOWNE
ISK
DIVERS

~ 303 rows ~

In all cases, you'll notice that a calculated value doesn't have a proper name.

Using Aliases

Calculated columns cause a minor inconvenience for SQL. Generally, each column should have a distinct name, but SQL has no clear idea what to call the newly generated column.

Some SQL clients will leave the calculated column unnamed, while some will generate a dummy name. When experimenting with simple SELECT statements, this is OK, but when taking the statement seriously, such as when you plan to use the results later, you will need to give each column a better name.

An **alias** is a new name for a column, whether it's a calculated column or an original. You create an alias using the AS keyword. For example:

```
SELECT
    id AS customer,
    height/2.54 AS height,
    givenname||' '||familyname AS fullname,
    --  givenname+' '+familyname AS fullname    --  MSSQL
    'active' AS status,
    (SELECT name FROM towns WHERE id=townid) AS town,
    length(email) AS length
    --  len(email) AS length           --  MSSQL
FROM customers;
```

This looks better:

customer	Height	fullname	status	town	length
42	66.339	May Knott	active	Kings Park	23
459	67.283	Rick Shaw	active	Richmond	24
597	60.236	Ike Andy	active	Hillcrest	23
186	69.291	Pat Downe	active	Guildford	24
352	61.575	Basil Isk	active	Wallaroo	24
576	69.409	Pearl Divers	active	Broadwater	27

~ 303 rows ~

Note

- The `id` column has been aliased even though it hasn't been calculated.

- The `height` calculation has been aliased to `height`; this is fine, since it still means the same thing, but in different units.

Apart from the fact that each calculated column must have a distinct name, other reasons to include aliases are as follows:

- Sometimes, you simply need to rename columns either for better meaning or to suit later use.

- Sometimes, you need to format or convert a column to something more suitable, but still retain its original name.

At this point, we're not worried about whether the preceding aliases are the best possible names for their columns; we're just looking at how they work.

Alias Names

By and large, the rules for alias names are the same as those for the names of columns. That means

- Aliases and original column names must be unique.

- Aliases should not contain spaces, can't start with a number, and can't contain other special characters.

- Aliases should not be SQL keywords.

If you really need to work around the preceding second and third rules, you can enclose the alias in double quotes. For example:

```
SELECT
    ordered AS "order",
    shipped AS "shipped date"
FROM sales;
```

Here, the name order is an SQL keyword, while shipped date contains a space.

Order	shipped date
2022-05-15 21:12:07.988741	2022-05-23
2022-05-16 03:03:16.065969	2022-05-24
2022-05-16 10:09:13.674823	2022-05-22
2022-05-16 15:02:43.285565	[NULL]
2022-05-16 16:48:14.674202	2022-05-28
~ 5549 rows ~	

You should resist the urge to do this. Aliases, as with column names, are for technical rather than aesthetic use. A SELECT statement is not actually a report.

Some DBMSs offer alternatives to double quotes for special names:

- Microsoft SQL offers square brackets as an alternative: [shipped date]. There is no reason to prefer this to double quotes.

- MySQL/MariaDB uses "backticks" as an alternative: 'shipped date'. In ANSI mode, this is unnecessary, but in traditional mode, it's all you've got.

Whatever names you choose, remember that they are meant to be purely functional. Don't get carried away trying to use upper and lower case, or spaces, or anything else that might look better. That's up to the software handling the output of your queries. In SQL, you just need a suitable name to refer to the data.

AS Is Optional

You will discover soon enough that AS is optional:

```
SELECT
    id customer,
    height/2.54 height,
    givenname||' '||familyname fullname,
    --  givenname+' '+familyname fullname    --  MSSQL
    'active' status,
```

110

```
    (SELECT name FROM towns WHERE id=townid) town,
    length(email) length
    --  len(email) length            --   MSSQL
FROM customers;
```

Some developers justify leaving out the AS as it saves time or makes them look more professional. However, you will also make this kind of mistake soon enough:

```
SELECT
    id,
    email
    givenname, familyname,
    height,
    dob
FROM customers;
```

This gives you a confusing result:

id	givenname	familyname	height	dob
42	may.knott61@example.net	Knott	168.5	[NULL]
459	rick.shaw459@example.net	Shaw	170.9	1945-07-03
597	ike.andy597@example.com	Andy	153	1998-08-09
186	pat.downe186@example.net	Downe	176	1990-04-12
352	basil.isk352@example.net	Isk	156.4	1960-01-13
576	pearl.divers576@example.com	Divers	176.3	

~ 303 rows ~

At first glance, this looks OK, as it is not a technical error. However, on closer inspection, you'll see that the email has been aliased to familyname, since there is no comma between them. Aliasing one column to another is legitimate, though it's not often that you would really want to.

You can't stop SQL from allowing this, but you can make mistakes like this slightly easier to spot if you develop a pattern which always includes AS for aliases.

Aliases Are Not Available in the Rest of the Query

Recall that the processing order for basic SQL clauses is

1. FROM

2. WHERE

3. GROUP BY

4. HAVING

5. SELECT

6. ORDER BY

This is different to the way you write SQL in that you *write* the SELECT clause first. This creates a major point of confusion in a statement like this:

```
SELECT id, title, price, price*0.1 AS tax
FROM books
WHERE tax<1.5;
```

This will result in an error, since although the price*0.1 AS tax expression is written in the first clause, it isn't actually processed until after the WHERE clause. As a result, tax is not yet available for the WHERE clause.

It becomes more confusing if you alias a calculation to an original column name:

```
SELECT
    id, title,
    price*1.1 AS price  --  adjust to include tax
FROM books
WHERE price<15;
```

This will work. Here, the price has been increased to include tax and aliased to the original name, which is legitimate.

id	Truncate	price
2078	The Duel	13.75
1530	Robin Hood, The Prince of Thieves	13.75
982	Struwwelpeter: Fearful Stories and Vile Pictures …	12.65
573	The Nose	11
1573	Rachel Ray	11
532	Elective Affinities	12.65
~ 521 rows ~		

However, the WHERE clause will filter on the *original* price column, *not* the adjusted version.

Again, there's not much you can do about this directly, as you don't have the option to write the SELECT clause further down, and you can't create aliases in any other clause.

Later, we will see how using Common Table Expressions can help preprocess calculated columns.

It's probably not a good idea to alias a calculation to an original column name *if* you're planning to use it later.

SQL has a clear idea of what it's going to do with the aliased name, but the human reader may well get confused.

Dealing with NULLs

Sooner or later, you will encounter the dreaded NULL in your calculations. Actually, there's nothing wrong with NULL *per se*, but it will totally make a mess of your calculations.

Any calculation involving a NULL will end up as NULL. You can say that NULL is very destructive for calculations. That is, unless you run the data through one or two expressions which are capable of handling NULLs.

The exception to this is with Oracle strings. Oracle treats NULL strings as empty strings. On one hand, this can be convenient; on the other, there are times when you really need NULL to behave as NULL.

If you are calculating with a single column which includes NULL, it makes sense that the result is also a NULL. For example:

```
SELECT
    id, givenname, familyname,
    height/2.54 AS height        -- sometimes NULL
FROM customers;
```

Sometimes, you get a NULL result:

id	Givenname	familyname	height
101	Artie	Chokes	63.858
489	Justin	Case	[NULL]
59	Leigh	Don	66.693
593	Luke	Warm	[NULL]
170	Dan	Dee	65.039
541	Neil	Downe	64.606

~ 303 rows ~

Since all we're doing is converting a single value, it is perfectly acceptable to leave NULLs as they are—if you don't know what the height is in centimeters, then you still don't know what it is in inches. However, we'll see shortly how you might sometimes replace the NULL with something you feel is better.

On the other hand, this behavior becomes more of a nuisance if you're calculating on multiple columns, most of which are not NULL:

```
SELECT
    id, givenname, othernames, familyname,
    givenname||' '||othernames||' '||familyname AS fullname
    -- MSSQL:
    -- givenname+' '+othernames+' '+familyname AS fullname
FROM authors;
```

With the exception of Oracle, you'll get a lot of NULLs:

id	givenname	Othernames	familyname	fullname
464	Ambrose	[NULL]	Bierce	[NULL]
858	Alexander	[NULL]	Ostrovsky	[NULL]
525	Francis	[NULL]	Beaumont	[NULL]
479	C.E.	Van	Koetsveld	C.E. van Koetsveld
703	Friedrich	[NULL]	Engels	[NULL]
~ 488 rows ~				

In the preceding example, most authors don't have a value for othernames, so it is NULL. Some don't even have a givenname value. There's nothing wrong for the most part with the givenname or the familyname, but the NULL for othernames destroys the whole calculation.

With Oracle, however, you won't get NULLs. However, you will see extra spaces where the missing names are.

Technically, the result is correct. If you don't know some of the names, then you don't know the full name. However, that's unhelpful.

Coalesce

SQL has a function called coalesce() which can replace NULL with a preferred alternative. The word "coalesce" actually means to combine, but how it came to be the name of this operation is one of those mysteries lost in the depths of ancient history.

The function is used this way:

```
coalesce(expression,planB)
```

If the preceding expression happens to be NULL, then planB will be used.

You can have a number of alternative values:

```
coalesce(expression,planB,planC, ... , planZ)
```

If planB also happens to be NULL, coalesce() will try the next alternative and so on until either there is a real value, or the alternatives have been exhausted.

You can see `coalesce` in action with missing phone numbers:

```
SELECT
    id, givenname, familyname,
    phone
FROM employees;
```

In the employees table, some are missing phone numbers.

id	Givenname	familyname	phone
7	Ebenezer	Splodge	0491577644
4	Gladys	Raggs	0491573087
28	Cornelius	Eversoe	[NULL]
32	Clarisse	Cringinghut	0491571804
33	Will	Power	0491576398
26	Fred	Kite	0491572983
~ 34 rows ~			

It would be reasonable to replace these missing phone numbers with the company's main phone number:

```
SELECT
    id, givenname, familyname,
    coalesce(phone,'1300975711')  -- coalesce to main number
FROM employees;
```

Here, the missing number has been coalesced:

id	Givenname	familyname	coalesce
7	Ebenezer	Splodge	0491577644
4	Gladys	Raggs	0491573087
28	Cornelius	Eversoe	1300975711
32	Clarisse	Cringinghut	0491571804

id	Givenname	familyname	coalesce
33	Will	Power	0491576398
26	Fred	Kite	0491572983
~ 34 rows ~			

The thing about coalesce() is that you can't always get away with it. You need to be sure that your substitute makes sense and that your guess is a good one. There are many times when it wouldn't make sense, such as a missing price for a book or an author's date of birth; NULL is often the best thing you can do.

In Chapter 2, you guessed at a missing quantity using coalesce() and then fixed it so that the quantity can't be NULL in the future. Sometimes, that's the best solution.

Fixing the Author Names

With the coalesce() function, you can replace the missing author names with an alternative. There are two things to consider here:

- You can't make up a missing name, so the alternative will have to be an empty string.

- You'll also want to leave out the spaces after the missing names.

For the second point, we won't coalesce just the name, but the combination of the name and the space, which should also be NULL—except for Oracle, which doesn't behave the same way. We'll take a different approach for Oracle.

To coalesce the names and spaces to an empty string, we can use

```
--  PostgreSQL, MariaDB/MySQL, SQLite
SELECT
    id, givenname, othernames, familyname,
    coalesce(givenname||' ','')
        ||coalesce(othernames||' ','')
        ||familyname AS fullname
FROM authors;

--  MSSQL
SELECT
```

117

```
    id, givenname, othernames, familyname,
    coalesce(givenname+' ',''')
        +coalesce(othernames+' ',''')
        +familyname AS fullname
FROM authors;
```

This gives us

id	givenname	Othernames	familyname	fullname
464	Ambrose	[NULL]	Bierce	Ambrose Bierce
858	Alexander	[NULL]	Ostrovsky	Alexander Ostrovsky
525	Francis	[NULL]	Beaumont	Francis Beaumont
479	C.E.	Van	Koetsveld	C.E. van Koetsveld
703	Friedrich	[NULL]	Engels	Friedrich Engels
~ 488 rows ~				

Since Oracle will happily concatenate a NULL string, we can't use coalesce().
Instead, we'll use the ltrim() function. This function removes leading spaces from a
string. Since we're adding a space to the *end* of the string, it will only be a leading space if
the name is empty. This gives us

```
-- Oracle
SELECT
    id, givenname, othernames, familyname,
    ltrim(givenname||' ')||ltrim(othernames||' ')
        ||familyname AS fullname
FROM authors;
```

This should give us the same result as before.

Using Calculations in Other Clauses

In this chapter, we're using the calculations mostly in the SELECT clause. Of course,
any clause which contains a value can use a calculated value. We'll have a look at a few
examples here.

One obvious use for calculations is in the WHERE clause. For example, you can find books with shorter titles:

```
SELECT *
FROM books
WHERE length(title)<24;      -- MSSQL: len(title)
```

giving

id	authorid	Title	published	price
2078	765	The Duel	1811	12.50
503	128	Uncle Silas	1864	17.00
2007	99	North and South	1854	17.50
702	547	Jane Eyre	1847	17.50
1759	17	La Curée	1872	16.00
205	436	Shadow: A Parable	[NULL]	17.50

~ 762 rows ~

You may need this if your database is case sensitive and you need to match a string in an unknown case:

```
SELECT *
FROM books
WHERE lower(title) LIKE '%journey%';
```

giving

id	authorid	Title	published	price
880	777	The Journey of Niels Klim to the Wor ...	1741	12.50
946	704	Following the Equator: A Journey Aro ...	1897	19.50
1314	606	Mozart's Journey to Prague	1856	17.00
1092	295	A Journey to the Western Islands of ...	1775	14.50
502	[NULL]	Journey to the Center of the Earth	1864	15.50
1454	914	A Sentimental Journey	1768	13.50

You can also calculate an aggregate value in a subquery:

```
SELECT *
FROM customers
WHERE height<(SELECT avg(height) FROM customers);
```

giving

id	familyname	Givenname	...	height	...
42	Knott	May	...	168.5	...
597	Andy	Ike	...	153.0	...
352	Isk	Basil	...	156.4	...
526	Coming	Seymour	...	163.5	...
26	Twishes	Bess	...	164.6	...
91	North	June	...	164.5	...
~ 128 rows ~					

You can also use calculations in the ORDER BY clause, such as when you want to sort by the length of a title:

```
SELECT *
FROM books
ORDER BY length(title);    -- MSSQL: length(title)
```

which gives

id	authorid	title	published	price
385	971	She	1887	11.00
488	478	Mumu	1852	18.00
728	534	Emma	1815	10.00
1625	496	Lenz	1835	18.50
317	99	Ruth	1853	16.50
2140	17	Nana	1880	12.50
~ 1200 rows ~				

120

However, you're likely to want to select what you're sorting by, so it would make more sense to calculate the value in the SELECT clause and sort by the result:

```
SELECT id, authorid, title, length(title) AS len, published, price
FROM books
ORDER BY len;        -- MSSQL: length(title)
```

which is more informative:

id	authorid	Title	len	published	price
385	971	She	3	1887	11.00
488	478	Mumu	4	1852	18.00
728	534	Emma	4	1815	10.00
1625	496	Lenz	4	1835	18.50
317	99	Ruth	4	1853	16.50
2140	17	Nana	4	1880	12.50
~ 1200 rows ~					

Here's an interesting use for coalesce in the ORDER BY clause. Some DBMSs support NULLS FIRST or NULLS LAST to decide where to put the NULLs in the sort order. If your DBMS doesn't support it, you can coalesce the column to an extreme value. For example:

```
SELECT *
FROM customers
ORDER BY coalesce(height,0);     -- NULLS FIRST
SELECT *
FROM customers
ORDER BY coalesce(height,1000); -- NULLS LAST
```

By coalescing all of the NULLs to an extreme value, SQL will sort them to one end or the other accordingly.

As for the FROM clause, you'll need a calculation which generates a virtual table. That's usually going to be a view, a join, or even a subquery. A Common Table Expression, in this context, is like a subquery. We'll do more of that sort of thing later.

More Details on Calculations

SQL databases typically understand the three main types of data: numbers, strings, and dates. There are variations on these types, such as whether numbers include decimals, or the length of a string, or whether the date includes a time. There are also some other types, such as boolean values (limited to true or false) or binary data (sometimes called Binary Large Objects, or BLOBs for short), with varying degrees of support in different DBMSs.

Here, we'll look at some of the details of calculating with the main data types.

As a rule, a value comes in three forms:

- A **stored** value may come from a variable or a column.

- A value may be **calculated** or come from a built-in function.

- A **literal** value may be entered directly in code.

SQL, like most coding languages, needs some help with certain literal values to distinguish them from other code. Numeric literals are entered as they are (bare) because they obviously can't be anything else.

String or date literals, on the other hand, are wrapped in single quotes (' ... ') to mark them as such. That's so that SQL can distinguish between strings and other words, such as SQL keywords or table and column names.

The actual value of a string or date literal doesn't include the quotes. However, the quotes are required when writing them into the code.

For much of what follows, we'll be using literals for examples.

Casting

The cast() function is used to interpret a value as a different data type. Recall that SQL has three main data types: numbers, strings, and dates. You can use cast to do one of two things:

- You can try to cast from one main type to another.

  Casting to a string should be easy enough, but casting to another type requires that SQL know how to interpret the value. Different DBMSs react differently to failure.

- You can cast within a main type. For example, you can cast between integer and decimal numbers or between dates and datetimes.

  If you cast a decimal to an integer, a datetime to a date, or a string to a shorter string, you'll naturally lose precision. If you cast in the opposite direction, the extra precision will be filled with the equivalent of "nothing."

  If you do cast to a narrower type, it will probably work, but don't push your luck too hard. For example, casting the number 123.45 to a decimal(4,2) will fail because you haven't allowed enough digits; you'll get an overflow error.

For what follows, remember that SQLite doesn't have a date type, so that's one cast you won't have to worry about. Later, we'll have a quick look at the equivalent in SQLite.

Here are some examples of casting within types:

```
-- shorter dates & numbers
SELECT
    -- not SQLite:
    cast(ordered as date) AS ordered_date,
    cast(total AS integer) AS whole_dollars
FROM sales;
-- shorter strings
SELECT cast(title AS varchar(16)) AS short_title
FROM books;

-- broader dates & numbers
SELECT
-- SQLite: no date type
-- PostgreSQL, Oracle
    cast(dob as timestamp) as long_dob,
-- MariaDB / MySQL, MSSQL
    cast(dob as datetime) as long_dob,

    cast(height as decimal(5,2)) as long_height
FROM customers;
```

If you cast a string to a longer type, one of two things will happen. If you cast it to a CHAR (fixed length) type, the extra length will be padded with spaces. If you cast it to a VARCHAR type, the string will be unchanged. However, the string will be permitted to grow to a longer string.

Casting between types is a different matter. Most DBMSs will automatically cast to a string if necessary. For example:

```
-- Not MSSQL
SELECT id || ': ' || email
FROM customers;
```

You'll get something like this:

?column?
42: may.knott61@example.net
459: rick.shaw459@example.net
597: ike.andy597@example.com
186: pat.downe186@example.net
352: basil.isk352@example.net
576: pearl.divers576@example.com
~ 303 rows ~

As you see, MSSQL won't do this automatically, possibly due to a confusion with their concatenation operator (+). There you'll have to force the issue:

```
-- MSSQL
SELECT cast(id as varchar(5)) + ': ' + email
FROM customers;
```

You can do the same with dates, too. We'll do that with the customers' dates of birth, but we'll run into the complication of the fact that some dates of birth are missing. Using coalesce should do the job:

```
-- PostgreSQL, MariaDB/MySQL, SQlite
   SELECT
       id || ': ' || email
       || coalesce(' Born: ' || dob,'')
   FROM customers;
-- MSSQL
   SELECT
       cast(id as varchar(5)) + ': ' + email
       + coalesce(' Born: ' + cast(dob as char(10)),'')
   FROM customers;
-- Not Oracle
```

For SQLite, it wasn't much effort as we've stored the dates as a string anyway.

Here, we've coalesced the entire concatenated value ' Born: ' || dob. That's because we want to replace the whole expression with the empty string if the dob is missing. Concatenating with a NULL should result in a NULL.

For Oracle, you run again into the quirk of treating NULL strings as empty strings, so they won't coalesce. We can work around it using CASE:

```
-- Oracle
   SELECT
       id || ': ' || email
       || CASE
            WHEN dob IS NOT NULL THEN ' Born: ' || dob
          END
   FROM customers;
```

Basically, you can think of coalesce as a simplified CASE expression. With Oracle, you need to spell it out more.

One reason you might want to change data types is to mix them with other values, such as concatenating the preceding strings. We'll also see casting being used when we want to combine data from multiple tables or virtual tables, such as with joins and unions.

Another reason to change data types is for sorting. All string data will normally sort alphabetically, but you may need to cast them as non-strings for sorting. For example:

```
--  Integers
    SELECT * FROM sorting
    ORDER BY numberstring;
    SELECT * FROM sorting
    ORDER BY cast(numberstring as int);            -- not MySQL
    --  ORDER BY cast(numberstring as signed);  --  MySQL

--  Dates (not SQLite)
    SELECT * FROM sorting
    ORDER BY datestring;
    SELECT * FROM sorting
    ORDER BY cast(datestring as date);
```

In the sorting table, there are some values stored as strings which represent numbers or dates. The only way to sort them properly is to cast them first.

Note that MySQL won't let you cast to an integer directly. You have to use SIGNED (which means the same thing) or UNSIGNED. MariaDB is OK with integers.

Not all casts from strings are successful, since the string may not resemble the correct type. For example:

```
--  This works:
    SELECT cast('23' as int)    --  MySQL: as signed
    --  FROM dual    --  Oracle
    ;

--  This doesn't:
    SELECT cast('hello' as int) --  MySQL: as signed
    --  FROM dual    --  Oracle
    ;
```

What happens next depends on the DBMS:

- MariaDB/MySQL will both give a 0 which is forgiving.

- MSSQL will give an error.

  However, you can use an alternative called try_cast which will simply give a NULL. If you wish, you can then coalesce the result.

- Oracle will also give an error.

  However, there is an optional default in this form:

  `cast('hello' as int DEFAULT 0 ON CONVERSION ERROR).`

  It's verbose but it allows an alternative to 0, or whatever you like.

- PostgreSQL just gives an error.

  It's possible to write a function to get around that.

Numeric Calculations

A number is normally used to count something—it's the answer to the question "how many." For example, how many centimeters in the customer's height, or how many dollars were paid for this item?

Numbers aren't always used that way. Sometimes, they're used as tokens or as codes. The calculations you might perform on a number would depend on how the number is being used.

Basic Arithmetic

You can always perform the basic operations on numbers:

```
SELECT
    3*5 AS multiplication,
    4+7 AS addition,
    8-11 AS subtraction,
    20/3 AS division,
    20%3 AS remainder,  -- Oracle: mod(20,3),
    24/3*5 AS associativity,
    1+2*3 AS precedence,
    2*(3+4) + 5*(8-5) AS distributive
-- FROM dual   -- Oracle
;
```

This sample illustrates the main operations:

mul...	add...	sub...	div...	rem...	ass...	pre...	dis...
15	11	-3	6	2	40	7	29

Note that you'll need to add FROM dual if you're testing this in Oracle. Also note

- Different DBMSs have different attitudes to dividing integers. In some cases, 20/3 would give you a result of 6, discarding the fraction. On other cases, you'd get something like 6.66...7 as a decimal.

- The % operator calculates the **remainder** after integer division. Oracle uses the mod() function.

- When mixing operations, SQL follows the rules you would have learned in school regarding precedence (which operators come first) and associativity (calculating from left to right). SQL also allows you to use parentheses to calculate expressions first.

If you know someone who's forgotten the basic rules of arithmetic, you can tell them

1. Do what's inside parentheses first.

2. Do multiplication | division before addition | subtraction (precedence).

3. Do operations of the same precedence from left to right (associativity).

Of course, these expressions work just the same whether the value is a literal or some stored or calculated value.

Mathematical Functions

There are some mathematical functions as well. For the most part, the mathematical functions won't get a lot of use unless you're doing something fairly specialized.

```
SELECT
    pi() AS pi,                 --  Not Oracle
    sin(radians(45)) AS sin45,  --  Not Oracle
    sqrt(2) AS root2,           --  √2
    log10(3) AS log3,
    ln(10) AS ln10,             --  Natural Logarithm
    power(4,3) AS four_cubed    --  4³
--  FROM dual                   --  Oracle
;
```

```
-- Oracle's Trigometric functions are less convenient
SELECT
    acos(-1) AS pi,
    sin(45*acos(-1)/180) AS sin45
FROM dual;
```

The results look something like this:

pi	sin45	root2	log3	ln10	four_cubed
3.142	0.707	1.414	0.477	2.303	64

So, now you can use SQL to find the length of a ladder leaning against a wall or the distance between two ships lost at sea.

Approximation Functions

There are also functions which give an *approximate* value of a decimal number. Here is a sample with variations between DBMSs:

```
SELECT
    ceiling(200/7.0) AS ceiling,
--  SQLite: round(200/7.0 + 0.5),
--  Oracle: ceil(200/7.0),

    floor(200/7.0) AS floor,
--  SQLite: round(200/7.0 - 0.5),

    round(200/7.0,0) AS rounded_integer,
--  or round(200/7), --  not MSSQL
    round(200/7.0,2) AS rounded_decimal

--  FROM DUAL   -- Oracle
;
```

As you see, the functions all tend to lose precision:

ceiling	floor	rounded_integer	rounded_decimal
29	28	29	28.57

If you use the cast() function to another narrow number type, you'll also lose precision. However, what happens next depends on the DBMS:

```
SELECT
    cast(234.567 AS int) AS castint,
    -- cast(234.567 AS unsigned),  --  MySQL
    cast(234.567 AS decimal(5,2)) AS castdec
-- FROM dual              -- Oracle
;
```

DBMS	Castint	castdec
PostgreSQL	235	234.57
MariaDB/MySQL	235	234.57
Oracle	235	234.57
MSSQL	234	234.57
SQLite	234	234.567

- With PostgreSQL, Oracle, and MariaDB/MySQL, casting to an integer or a shorter decimal will round off the number.

- With MSSQL, casting to a shorter decimal will round off the number, but casting to an integer will truncate it. If you want the integer truncated, you can use something like decimal(3,0).

- With SQLite, casting to an integer will truncate, while casting to a decimal is ignored and retains the original value.

Formatting Numbers

Formatting functions change the *appearance* of a number. Unlike approximation and other functions, the result of a formatting function is not a number but is a string; that's the only way you can change the way a number appears.

For numbers, most of what you want to do is change the number of decimal places, display the thousands separator, and possibly currency symbols.

Again, the different DBMSs have wildly different functions. As an example, here are some ways of formatting a number as currency with thousands separators:

```
-- PostgreSQL, Oracle
   SELECT
       to_char(total,'FM999G999G999D00') AS local_number,
       to_char(total,'FML999G999G999D00') AS local_currency
   FROM sales;
   SELECT to_char(total,'FM$999,999,999.00') FROM sales;

-- MariaDB/MySQL
   SELECT
           format(total,2) AS local_number,
           format(total,2,'de_DE') AS specific_number
   FROM sales;

-- MSSQL
   SELECT
       format(total,'n') AS local_number,
       format(total,'c') AS local_currency
   FROM sales;

-- SQLite
   SELECT printf('$%,d.%02d',total,round(total*100)%100)
   FROM sales;
```

You'll get variations of the following:

local_number	local_currency
28.00	$28.00
34.00	$34.00
58.50	$58.50
50.00	$50.00
17.50	$17.50
13.00	$13.00
~ 5549 rows ~	

Note

- Both PostgreSQL and Oracle have a flexible `to_char()` function which can also be used to format dates.

- MariaDB/MySQL uses the `format()` function which adds thousands separators and decimal places; you can also tell it to adjust for different locales.

- MSSQL has its own `format()` function with its more intuitive formatting codes; it also adjusts for locales and can be used to format a date.

- SQLite only has a generic `format()`, a.k.a. `printf()`, function, which will be more familiar to programmers; SQLite presumes that you will format data in the host application such as PHP or wherever SQLite has been embedded.

Note that if you do run a number through a formatting function, *it is no longer a number!* If all you do is look at it, then that doesn't matter. However, if you have plans to do any further calculations, or to sort the results, then a formatted number is likely to backfire on you.

When all is said and done, formatting is probably something you won't do much in SQL. The main purpose of SQL is to *get* the data and prepare it for the next step. Formatting comes last and is often done in other software.

String Calculations

A **string** is a string of characters, hence the name. In SQL, this is referred to as **character** data.

Traditionally, SQL has two main data types for strings:

- Character: `CHAR(length)` is a fixed-length string. If you enter fewer characters than the length, then the string will be right-padded with spaces. This probably explains why standard SQL ignores trailing spaces for string comparison.

- Character varying: `VARCHAR(length)` is a limited length string. If you enter a shorter string, it will *not* be padded.

In principle, CHAR() is more efficient for processing since it's always the same length, and the DBMS doesn't need to worry about working out the size and making things fixed. VARCHAR() is supposed to be more efficient for storage.

In reality, modern DBMSs are much cleverer than their ancestors, and the difference between the two types is not very important anymore. For example, PostgreSQL recommends always using VARCHAR since it actually handles that type more efficiently.

Most DBMSs offer a third type, TEXT, which is, in principle, unlimited in length. Again, modern DBMSs allow longer standard strings than they used to, so again this is not so important. Microsoft has deprecated TEXT in favor of VARCHAR(MAX) which does the same job.

A string literal is written between single quotes:

```
SELECT 'hello'; --  Oracle: FROM dual;
```

When working with strings, you normally simply want to save them and fetch them. However, you can process the strings themselves. This is usually one of the following operations:

- **Concatenation** means joining strings together.

  Concatenation is the only direct operation on strings. All other operations make use of functions.

- Some functions will make changes to a string. They don't actually change the string, but return a changed version of the string.

- Some functions can be used to extract parts of a string.

- Some functions are more concerned with individual characters of the string.

Case Sensitivity

SQL will store the upper/lower case characters as expected, but you may have a hard time searching for them. That's because some databases ignore case, while others don't.

How a database handles case is a question of **collation**. Collation refers to how it interprets variations of letters. In English, the only variation to worry about is upper or lower case, but other languages may have more variations, such as accented letters in French or German.

Collation will have an impact on how strings are sorted and how they compare. In English, you're mainly worried about whether upper case strings match lower case strings and possibly whether upper and lower case strings are sorted together or sorted separately. In some other languages, the same questions might apply to whether accented and nonaccented characters match and how they, too, are sorted.

You can set a collation when you create the database or a table, but if you don't worry about it, the DBMS will have a default collation for new databases.

In PostgreSQL, Oracle, and SQLite, the default collation is case sensitive, so upper and lower case won't match. With MySQL/MariaDB and MSSQL, the default collation is case insensitive, so they will match.

If you're not sure whether your particular database is case sensitive or not, you can try this simple test:

```
SELECT * FROM customers WHERE 'a'='A';
```

If the database is case sensitive, you won't get any rows, since a won't match A; if it's not, you will get the whole table.

ASCII and Unicode

Traditionally, strings are encoded using ASCII—the American Standard Code for Information Interchange. Each character has a number from 32 to 126, stored in a single byte. For example, A is encoded as 65, while a is encoded as 97. Special characters include the space (32), the exclamation mark (33), and even the numerals 0–9 (48–57).

Since there are only 95 values between 32 and 126, ASCII has a limited range of characters. Once you've taken up the alphabet in upper and lower case as well as the 10 digits, you've already used up 62 characters, which doesn't leave much room for punctuation or other special characters. (Why they include obscure characters such as ~ and ` remains a mystery.)

Basic ASCII certainly doesn't have the scope to include more punctuation characters, European accented characters, or the Greek or Cyrillic alphabets. And don't even think about Japanese or Chinese.

One technique for handling other languages is to switch to different variations of ASCII. A more enduring solution is to use Unicode.

Unicode is the modern standard for handling multiple languages in a single encoding system. It does this by using multiple bytes. How exactly this is achieved can be tricky and will vary in different implementations, but the idea is the same.

Unicode is designed to include ASCII codes, so there is some compatibility between the two. However, some Unicode implementations do take up more space than ASCII, even when they're encoding the same characters. Today, space is cheap, and database software is pretty clever at using space efficiently, so that shouldn't be too much of a problem.

All modern DBMSs support Unicode. Some do it by default, while some expect you to ask for it. In some cases, you can use Unicode for the whole database, for particular tables, or for individual columns.

The sample database uses Unicode for most of the data, but may use ASCII in some cases where the character set is deliberately limited, such as for phone numbers which are stored as strings.

Some DBMSs support NCHAR and NVARCHAR data types in addition to CHAR and VARCHAR. If the database tables are set to use Unicode, then CHAR and VARCHAR will do the job. Otherwise, you might use NCHAR and NVARCHAR to specify Unicode on particular columns.

Concatenation

Concatenation means joining strings together. This is the simplest string operation and the only one which can be done without a function.

The concatenation operator is usually ||. Microsoft SQL Server uses + instead. For example:

```
SELECT
    id,
    givenname||' '||familyname AS fullname
    --  givenname+' '+familyname AS fullname    --  MSSQL
FROM customers;
```

That will give you something like

Id	fullname
42	May Knott
459	Rick Shaw
597	Ike Andy
186	Pat Downe
352	Basil Isk
576	Pearl Divers

~ 303 rows ~

Note that MySQL in traditional mode doesn't support the concatenation operator in any form. In ANSI mode, it supports the standard || operator.

Many DBMSs also support a non-standard function concat(string,string,...). For example:

```
-- Not SQLite
SELECT
    id,
    concat(givenname,' ',familyname) AS fullname
FROM customers;
```

This is not supported in SQLite. However, it is supported in MySQL, so that's how you concatenate strings in traditional mode.

For most DBMSs, there is a subtle but important difference between the concat() function and the concatenation operator. With the concatenation operator, if there is a NULL in the mix, the result will (naturally) also be NULL. However, the concat() function will automatically coalesce a NULL result to an empty string (''). This may or may not be convenient, as sometimes the NULL is something you should know about.

Oracle, however, takes a different approach. They regard a NULL string as the same as an empty string '', so concatenating a NULL either way is the same as concatenating an empty string. On one hand, this is convenient if you don't want to have to coalesce; on the other hand, there are times when you need NULL to be NULL, so this can be awkward.

String Functions

Other operations with strings require functions. Here are some examples.

For the following examples, we've included SELECT * for context—except that in Oracle you need to write SELECT table.* if you're mixing it with other data, so we've done that with all of the examples which include Oracle.

The length of a string is the number of characters in the string. To find the length, you can use

```
--  PostgreSQL, MySQL/MariaDB, SQLite, Oracle
    SELECT customers.*, length(familyname) AS len
    FROM customers;
--  MSSQL
    SELECT *, len(familyname) AS len FROM customers;
```

To find where part of a string is, you can use the following:

```
--  MySQL/MariaDB, SQLite, Oracle: INSTR('values',value)
    SELECT books.*, instr(title,' ') AS space FROM books;

--  PostgreSQL: POSITION(value IN 'values')
    SELECT *, position(' ' in title) AS space FROM books;

--  MSSQL: CHARINDEX(value, 'values')
    SELECT *, charindex(' ',title) AS space FROM books;
```

You can use replace to replace substrings in a string:

```
--  replace(original,search,replace)
    SELECT books.*, replace(title,' ','-') AS hyphens
    FROM books;
```

To change between upper and lower case, there is

```
--  PostgreSQL, MySQL/MariaDB, SQLite, Oracle, MSSQL
    SELECT
        books.*,
        upper(title) AS upper,
```

```
        lower(title) AS lower
    FROM books;
```

```
-- PostgreSQL, Oracle
    SELECT books.*, initcap(title) AS lower FROM books;
```

To remove extra spaces at the beginning or the end of a string, you can use trim() to remove from both ends, or ltrim() or rtrim() to remove from the beginning or end of the string:

```
WITH vars AS (
    SELECT ' abcdefghijklmnop ' AS string
    -- FROM dual    -- Oracle
)
SELECT
    string,
    ltrim(string) AS ltrim,
    rtrim(string) AS rtrim,
    trim(string) AS trim AS trim,
    ltrim(rtrim(string)) AS same
FROM vars;
```

All modern DBMSs support trim(), but MSSQL didn't until version 2017. PostgreSQL also calls it btrim(). You may not notice when the spaces on the right are trimmed.

You can get substring with substring() or substr(), depending on your DBMS:

```
WITH vars AS (
    SELECT 'abcdefghijklmnop' AS string
    FROM dual    -- Oracle
)
SELECT
-- PostgreSQL, MariaDB/MySQL, Oracle, SQLite
    substr(string,3,5) AS substr,
-- PostgreSQL, MariaDB/MySQL, MSSQL, SQLite
    substring('abcdefghijklmnop',3,5) AS substring
FROM vars;
```

Some DBMSs include specialized functions to get the first or last part of a string. In some cases, you can use a negative start to get the last part of a string:

```
WITH vars AS (
    SELECT 'abcdefghijklmnop' AS string
    FROM dual    --  Oracle
)
SELECT
-- Left
    --  PostgreSQL, MariaDB/MySQL, MSSQL:
        left('abcdefghijklmnop',4) AS lstring
    --  All DBMSs including SQLITE and Oracle:
    --  substr(string,1,n) AS lstring,
-- Right
    --  PostgreSQL, MariaDB/MySQL, MSSQL:
        right('abcdefghijklmnop',4) AS rstring
    --  MariaDB/MySQL, Oracle, SQLite
    --  substr('abcdefghijklmnop',-4) AS rstring
FROM vars;
```

Just note that if you spend a lot of time extracting substrings from your data, it's possible that you're trying to store too much in a single value.

On the other hand, you can often use substrings to reformat raw data into something more friendly.

Date Operations

From an SQL point of view, dates are problematic. That's because, despite their overwhelming presence in daily life, measuring dates is a mess.

One problem is that we measure dates using a number of incompatible cycles all at the same time: the day, week, month, and year. To make things worse, we all live in different time zones, so we can't even agree on what time it is.

Most DBMSs have a number of related data types to manage dates, specifically the date which is for dates with times and datetime which includes the time. Generally, you can expect variations on these types, as well as the ability to include time zones.

The exception is SQLite, which expects you to use numbers or strings and run the values through a few functions to do the date arithmetic.

There are a number of things you would expect to do with dates and times:

1. Enter and store a date/time

2. Get the current date/time

3. Group and sort by date/time

4. Extract parts of the date/time

5. Add to a date/time

6. Calculate the difference between two dates/times

7. Format a date/time

SQLite has a completely different approach to working with dates. That's partly because it doesn't actually support dates. As a result, SQLite will be missing from much of the following discussion. The Appendix has some information on handling dates in SQLite.

Entering and Storing a Date/Time

Since most DBMSs have their own way of storing a date/time, the actual details of date storage are not important. What is important is that you can enter the data.

In a table, a date or datetime column is usually defined as follows:

DBMS	Date	Date with time
PostgreSQL	DATE	TIMESTAMP
MariaDB/MySQL	DATE	DATETIME
MSSQL	DATE	DATETIME2
Oracle	DATE	DATETIME
SQLite	TEXT	TEXT

The normal way to enter a `date` or `datetime` literal is to use one of the following:

- date: `'2013-02-15'`

- datetime: `'2013-02-15 09:20:00'`

You can also omit the seconds or include decimal parts of a second.

The format is a variation of the **ISO8601** format. In pure ISO8601 format, the time would be written after a T instead of a space.

Note that with Oracle, datetime literals generally use a different format. To use the preceding formats, prefix the literal with `date` or `datetime`, respectively:

- date: `date '2013-02-15'`

- datetime: `datetime '2013-02-15 09:20:00'`

In PostgreSQL, MSSQL, and MySQL/MariaDB, you can often enter another readable date format such as `'15 Feb 2013'`. However, you should *never* use the format `'2/3/2013'` which has different meanings internationally.

In practical terms, just stick to the standard format:

```
SELECT *
FROM customers
WHERE dob<'1980-01-01'; --  Oracle dob<date '1980-01-01';
```

which gives you older customers:

id	givenname	familyname	...	dob	...
459	Rick	Shaw	...	1945-07-03	...
352	Basil	Isk	...	1960-01-13	...
92	Nan	Keen	...	1943-05-18	...
267	Boris	Todeath	...	1969-10-06	...
91	June	North	...	1967-03-22	...
543	Nat	Ering	...	1946-04-30	...
~ 133 rows ~					

Note that in simple expressions like `dob<'1980-01-01'`, SQL doesn't get confused about whether the expression is a date or a string: the context makes it clear.

Getting the Current Date/Time

One thing you will want to do is compare a date/time to now. In most DBMSs, you can use

```
SELECT
    current_timestamp AS now,
    current_date AS today        --   Not MSSQL
-- FROM dual   --  Oracles
;
```

Note

- MSSQL also has `getdate()` as a synonym for `current_timestamp`. Despite the name, it gives not just the date.

- MariaDB/MySQL also has `now()` as a synonym for `current_timestamp`.

- Oracle also has `systemtimestamp` and `systemdate` for date/time on the database server rather than on the client.

As noted earlier, MSSQL doesn't have a version of `current_date`. In any case, you may have an existing `datetime` which you want to simplify to a `date`. The simplest way is to `cast` the `datetime`:

```
--  Not Oracle
    SELECT
        current_timestamp AS now,
        cast(current_timestamp as date) AS today
    --  FROM dual   --  Oracle
    ;
```

This won't quite work with Oracle; it will let you do the cast all right, but it doesn't change anything. Instead, you should use the `trunc()` function:

```
--  Oracle
    SELECT
        current_timestamp AS now,
```

```
    trunc(current_timestamp) AS today
FROM dual    --  Oracle
;
```

This will still have a time component, but it's set to 00:00.

Grouping and Sorting by Date/Time

You can sort by date/time as with any other data type. The result will be in historical order:

```
SELECT *
FROM sales
ORDER BY ordered;
```

Of course, you can also use DESC.

You can also group by date, but you probably wouldn't want to group by datetime, unless you have a huge number of transactions per second. For a datetime, you might use a Common Table Expression to cast it to a date and then group the results. For example:

```
WITH cte AS (
    SELECT
        cast(ordered as date) AS ordered, total  --  Not Oracle
        -- trunc(ordered) AS ordered, total      --  Oracle
    FROM sales
)
SELECT ordered, sum(total)
FROM cte
GROUP BY ordered
ORDER BY ordered;
```

This gives you the following summary:

ordered	sum
2022-05-04	43.00
2022-05-05	150.50
2022-05-06	110.50
2022-05-07	142.00

ordered	sum
2022-05-08	214.50
2022-05-09	16.50
~ 389 rows ~	

Remember, in Oracle you need to use the `trunc()` function.

Extracting Parts of a Date/Time

Technically, a `datetime` represents a point in time. Practically, we tend to think in terms of components such as days and years. The situation is complicated by the fact that (a) the components are not in step with each other and (b) some of them vary in size.

Date Extracting in PostgreSQL, MariaDB/MySQL, and Oracle

The standard method of extracting part of a date is to use the `extract()` function. This function takes the form

```
extract(part from datetime)
```

You can see the `extract()` function in action:

```
WITH chelyabinsk AS (
    SELECT
        timestamp '2013-02-15 09:20:00' AS datetime
    FROM dual
)
SELECT
    datetime,
    EXTRACT(year FROM datetime) AS year,
    EXTRACT(month FROM datetime) AS month,
    EXTRACT(day FROM datetime) AS day,
    --   not Oracle or MariaDB/MySQL:
        EXTRACT(dow FROM datetime) AS weekday,
    EXTRACT(hour FROM datetime) AS hour,
```

```
    EXTRACT(minute FROM datetime) AS minute,
    EXTRACT(second FROM datetime) AS second
FROM chelyabinsk;
```

You get the following components:

datetime	year	month	day	weekday	hour	minute	second
2013-02-15 09:20:00	2013	2	15	5	9	20	0

Note that Oracle and MariaDB/MySQL don't have a direct way of extracting the day of the week, which can be a problem if, say, you want to use it for grouping. However, as you will see later, you can use a formatting function to get the day of the week, as well as the preceding values.

PostgreSQL also includes a function called date_part('part',datetime) as an alternative to the preceding function.

Date Extracting in Microsoft SQL

Microsoft SQL has two main functions to extract part of a date:

- datepart(part,datetime) extracts the part of a date/time as a *number*.

- datename(part,datetime) extracts the part of a date/time as a *string*. For most parts, such as the year, it's simply a string version of the datepart number. However, for the weekday and the month, it's actually the human-friendly name.

You can see these two functions in action:

```
WITH chelyabinsk AS (
    SELECT cast('2013-02-15 09:20' as datetime) AS datetime
)
SELECT
    datepart(year, datetime) AS year,      -- aka year()
    datename(year, datetime) AS yearstring,
    datepart(month, datetime) AS month,  -- aka month()
    datename(month, datetime) AS monthname,
    datepart(day, datetime) AS day,        -- aka day()
```

```
    datepart(weekday, datetime) AS weekday, --   Sunday=1
    datename(weekday, datetime) AS weekdayname,
    datepart(hour, datetime) AS hour,
    datepart(minute, datetime) AS minute,
    datepart(second, datetime) AS second
FROM chelyabinsk;
```

Note

- datename(date,year) just gives a string version of 2013.

- There are three short functions—day(), month(), and year()—which are synonyms of datepart().

Formatting a Date

As with numbers, formatting a date generates a string.

For both PostgreSQL and Oracle, you can use the to_char function. Here are two useful formats:

```
--  PostgreSQL
    WITH vars AS (SELECT timestamp '1969-07-20 20:17:40' AS moonshot)
    SELECT
        moonshot,
        to_char(moonshot,'FMDay, DDth FMMonth YYYY') AS fulldate,
        to_char(moonshot,'Dy DD Mon YYYY') AS shortdate
    FROM vars;
```

```
--  Oracle
    WITH vars AS (
        SELECT timestamp '1969-07-20 20:17:40' AS moonshot FROM dual
    )
    SELECT
        moonshot,
        to_char(moonshot,'FMDay, ddth Month YYYY') AS fulldate,
        to_char(moonshot,'Dy DD Mon YYYY') AS shortdate
    FROM vars;
```

146

You'll get something like this:

moonshot	full	short
1969-07-20 20:17:40	Sunday, 20th July 1969	Sun 20 Jul 1969

You'll notice that there is a slight difference in the format codes between PostgreSQL and Oracle.

For MariaDB/MySQL, there is the date_format() function:

```
WITH vars AS (SELECT timestamp '1969-07-20 20:17:40' AS moonshot)
SELECT
    moonshot,
    date_format(moonshot,'%W, %D %M %Y') AS fulldate,
    date_format(moonshot,'%a %d %b %Y') AS shortdate
FROM vars;
```

For Microsoft SQL, the format() function can also be used for dates:

```
WITH vars AS (SELECT cast('1969-07-20 20:17:40' AS datetime) AS moonshot)
SELECT
    format(moonshot,'dddd, d MMMM yyy') AS fulldate,
    format(moonshot,'ddd d MMM yyy') AS shortdate
FROM vars;
```

SQLite has very limited formatting functionality, and you certainly can't get month or weekday names without some additional trickery. It's usually better to leave the date alone and let the host application do what is needed.

You can learn more about the format codes at

- PostgreSQL: www.postgresql.org/docs/current/functions-formatting.html#FUNCTIONS-FORMATTING-DATETIME-TABLE

- Oracle: https://docs.oracle.com/en/database/oracle/oracle-database/21/sqlrf/Format-Models.html

- MariaDB: https://mariadb.com/kb/en/date_format/

- MySQL: https://dev.mysql.com/doc/refman/8.0/en/date-and-time-functions.html

- Microsoft SQL: `https://learn.microsoft.com/en-us/dotnet/standard/base-types/custom-date-and-time-format-strings`

Date Arithmetic

Generally, the two things you want to do with dates are

- Modify a date by adding or subtracting an interval

- Find the difference between two dates

To modify a date, you can add or subtract an interval. Some DBMSs define a type of data called `interval` for the purpose. For example, to add four months to now, you can use

```
--  PostgreSQL
SELECT
     date '2015-10-31' + interval '4 months' AS afterthen,
     current_timestamp + interval '4 months' AS afternow,
     current_timestamp + interval '4' month  --  also OK     ;

--  Oracle
SELECT
     add_months('31 Oct 2015',4) AS afterthen,
     current_timestamp + interval '4' month AS afternow,
     add_months(current_timestamp,4) --  also OK
FROM dual;

--  MariaDB/MySQL
SELECT
     date_add('2015-10-31',interval 4 month) AS afterthen,
     date_add(current_timestamp,interval 4 month)
         AS afternow,
     current_timestamp + interval '4' month  --  also OK
    ;
```

This gives you something like this:

afterthen	Afternow
2016-02-29 00:00:00	2023-10-01 16:01:13.691447+11

You'll notice that PostgreSQL and Oracle use the addition operator, while MariaDB/MySQL uses a special function. Oracle also has a special function to add months.

For Microsoft SQL, you use `dateadd`, specifying the units and number of units:

```
-- MSSQL
 SELECT
      dateadd(month,4,'2015-10-31') AS afterthen,
      dateadd(month,4,current_timestamp) AS afternow
 ;
```

SQLite uses the `strftime()` function to convert from a string, together with modifiers to adjust the date:

```
-- SQLite
 SELECT
      strftime('%Y-%m-%d','2015-10-31','+4 month')
          AS afterthen,
      strftime('%Y-%m-%d','now','+4 month') AS afternow
 ;
```

The other thing you'll want to do is calculate the difference between two dates. Here again, every DBMS does it differently. For example, to find the age of your customers, you can use

```
-- PostgreSQL
 SELECT
      dob,
      age(dob) AS interval,
      date_part('year',age(dob)) AS years,
      extract(year from age(dob)) AS samething
 FROM customers;

-- MariaDB/MySQL
 SELECT
      dob,
      timestampdiff(year,dob,current_timestamp) AS age
 FROM customers;
```

```
--  MSSQL, but not quite!
    SELECT
        dob,
        datediff(year,dob,current_timestamp) AS age
    FROM customers;

--  Oracle
    SELECT
        dob,
        trunc(months_between(current_timestamp,dob)/12)
            AS age
    FROM customers;
--  SQLite
    SELECT
        dob,
        cast(
            strftime('%Y.%m%d', 'now')
            - strftime('%Y.%m%d', dob)
        as int) AS age
    FROM customers;
```

For PostgreSQL, you'll get the following results. The other DBMSs won't have the age column:

dob	interval	Years	samething
[NULL]	[NULL]	0	0
1945-07-03	77 years 10 mons 29 days	77	77
1998-08-09	24 years 9 mons 23 days	24	24
1990-04-12	33 years 1 mon 19 days	33	33
1960-01-13	63 years 4 mons 19 days	63	63
[NULL]	[NULL]	0	0

~ 303 rows ~

Of the preceding calculations, MSSQL has a simple function which is *too* simple. All it does is calculate the difference between the years, which is way out if the date of birth is at the end of the year but the asking date is at the beginning of the year. To get a more correct result takes a lot more work.

The CASE Expression

There are times when a simple expression won't do, and you need SQL to make some choices. The CASE ... END expression can be used to choose from alternative values.

For example, you can create categories from other values:

```
SELECT
    id,title,
    CASE
        WHEN price<13 THEN 'cheap'
        WHEN price<=17 THEN 'reasonable'
        WHEN price>17 THEN 'expensive'
        -- ELSE NULL
    END AS price
FROM books;
```

You get a simple price listing:

id	Title	price
2094	The Manuscript Found in Saragossa	expensive
336	The Story of My Life	reasonable
1868	The Tenant of Wildfell Hall	[NULL]
375	Dead Souls	reasonable
1180	Fables	cheap
990	The History of Pendennis: His Fortun …	cheap

~ 1200 rows ~

Note that if all conditions fail, then the result will be NULL, which is commented out earlier. If you want an alternative to NULL, use the ELSE expression:

```
SELECT
    id,title,
    CASE
        WHEN price<13 THEN 'cheap'
        WHEN price<=17 THEN 'reasonable'
        WHEN price>17 THEN 'expensive'
        ELSE ''
    END AS price
FROM books;
```

Also, note that the CASE expression is **short-circuited**: once it finds a match, it stops evaluating.

Various Uses of CASE

There is a simplified variation of CASE for when you are testing a simple discrete value. For example:

```
SELECT
    c.id,
    givenname||' '||familyname AS name,
    -- givenname+' '+familyname AS name, --  MSSQL
    CASE status
        WHEN 1 THEN 'Gold'
        WHEN 2 THEN 'Silver'
        WHEN 3 THEN 'Bronze'
    CASE AS status
FROM customers AS c LEFT JOIN VIP ON c.id=vip.id;
-- Oracle:
--  FROM customers c LEFT JOIN VIP ON c.id=vip.id;
```

This gives you

id	Name	status
69	Rudi Mentary	[NULL]
182	June Hills	Bronze
43	Annie Day	[NULL]
263	Mark Time	Bronze
266	Vic Tory	Silver
68	Phyllis Stein	[NULL]
442	Herb Garden	Gold
33	Eileen Dover	[NULL]
~ 303 rows ~		

This form isn't much shorter, but it makes the intention clear.

You can also use the IN expression:

```
SELECT
    id, givenname, familyname,
    CASE
        WHEN state IN('QLD','NSW','VIC','TAS') THEN 'East'
        WHEN state IN ('NT','SA') THEN 'Central'
        ELSE 'Elsewhere'
    END AS region
FROM customerdetails;
```

which gives you

id	Givenname	familyname	region
137	Albert	Ross	East
359	Gail	Warning	Central
40	Cliff	Face	East
151	Rick	O'Shea	East

id	Givenname	familyname	region
96	Rob	Blind	Elsewhere
465	Mary	Christmas	Elsewhere

~ 303 rows ~

Coalesce Is like a Special Case of CASE

There's some similarity between using coalesce() and CASE. You can think of CASE as an alternative to coalesce:

```
SELECT
    id, givenname, familyname,
    coalesce(phone,'-') AS coalesced,
    CASE
        WHEN phone IS NOT NULL THEN phone
        ELSE '-'
    END AS cased
FROM customers;
```

The two expressions will give the same results:

id	givenname	familyname	coalesced	cased
42	May	Knott	0255509371	0255509371
459	Rick	Shaw	0370101040	0370101040
597	Ike	Andy	-	-
186	Pat	Downe	0870105900	0870105900
352	Basil	Isk	0255502503	0255502503
576	Pearl	Divers	0370107821	0370107821

~ 303 rows ~

It's not necessarily a convenient alternative, of course, but it helps to appreciate the overlapping use of the two. It's particularly useful with Oracle, where you can happily concatenate a NULL without ending up with a NULL, so it's hard to coalesce otherwise.

Nested CASE Expression

CASE can also be nested with additional CASEs. This is useful when there are possibilities within possibilities.

For example, the sales table has the date and time when the order was placed and the date when (or if) the order was shipped.

We can use CASE to generate a status for these dates. For example, using the shipped date and ordered date, you can set up the following criteria:

- Shipped: Compare shipped to ordered

    - 14 days ⇒ Shipped Late

    - Else Shipped

- Not Shipped: Compare Today to ordered

    - < 7 days ⇒ Current

    - < 14 days ⇒ Due

    - Else Overdue

Before we get going, however, note that some sales have no ordered value:

```
SELECT * FROM sales;
```

That might be, for example, if the customer never checked out the order. We probably should get rid of them, but, for now, we'll just filter them out:

```
SELECT * FROM sales WHERE ordered IS NOT NULL;
```

The first thing you'll have to do is to calculate the difference between dates. This varies between DBMSs:

```
-- PostgreSQL, MariaDB / MySQL, Oracle
    SELECT
        id, customerid, total,
        cast(ordered as date) AS ordered, shipped,
```

```
         current_date - cast(ordered as date) AS ordered_age,
         shipped - cast(ordered as date) AS shipped_age
    FROM sales
    WHERE ordered IS NOT NULL;
--   MSSQL
    SELECT
         id, customerid, total,
         cast(ordered as date) AS ordered, shipped,
         datediff(day,ordered,current_timestamp) AS ordered_age,
         datediff(day,ordered,shipped) AS shipped_age
    FROM sales
    WHERE ordered IS NOT NULL;
--   SQLite
    SELECT
         *,
         julianday('now')-julianday(ordered) AS ordered_age,
         julianday(shipped)-julianday(ordered) AS shipped_age
    FROM sales
    WHERE ordered IS NOT NULL;
```

You'll get the following:

id	customerid	total	ordered	shipped	ordered_age	shipped_age
39	28	28.00	2022-05-15	2022-05-23	382	8
40	27	34.00	2022-05-16	2022-05-24	381	8
42	1	58.50	2022-05-16	2022-05-22	381	6
43	26	50.00	2022-05-16	[NULL]	381	[NULL]
45	26	17.50	2022-05-16	2022-05-28	381	12
668	105	15.00	2022-07-27	[NULL]	309	[NULL]
~ 5295 rows ~						

Note that with SQLite, the simplest way to get an age is to convert dates to a Julian date, which is the number of days since Noon, 24 November 4714 BC. Long story.

You know by now that you can't use the calculated values in other parts of the SELECT clause, so that's awkward if you need them. You can, however, do the query in two steps.

If you put the preceding query in a Common Table Expression, you can then use the results in the main query.

First, you need to distinguish between those which have been shipped and those which haven't:

```
WITH salesdata AS (
    --  one of the above queries WITHOUT the semicolon
)
SELECT
    salesdata.*,
    CASE
        WHEN shipped IS NOT NULL THEN
            --  One of two statuses
        ELSE
            --  One of three statuses
    END AS status
FROM salesdata;
```

The statuses in each case are additional CASE expressions:

```
WITH salesdata AS (
    --  one of the above queries WITHOUT the semicolon
)
SELECT
    salesdata.*,
    CASE
        WHEN shipped IS NOT NULL THEN
            CASE
                WHEN shipped_age>14 THEN 'Shipped Late'
                ELSE 'Shipped'
            END
        ELSE
```

```
                    CASE
                        WHEN ordered_age<7 THEN 'Current'
                        WHEN ordered_age<14 THEN 'Due'
                        ELSE 'Overdue'
                    END
            END AS status
    FROM salesdata;
```

This will give you something like

id	cid	total	ordered	shipped	ordered_age	shipped_age	status
39	28	28.00	2022-05-15	2022-05-23	382	8	Shipped
40	27	34.00	2022-05-16	2022-05-24	381	8	Shipped
42	1	58.50	2022-05-16	2022-05-22	381	6	Shipped
43	26	50.00	2022-05-16	[NULL]	381	[NULL]	Overdue
45	26	17.50	2022-05-16	2022-05-28	381	12	Shipped
668	105	15.00	2022-07-27	[NULL]	309	[NULL]	Overdue

~ 5295 rows ~

Summary

Data in an SQL table should be stored in its purest, simplest form. However, this data can be recalculated to increase its usefulness.

Calculations can take a number of forms:

- Based on single columns

- Based on multiple columns

- Hard-coded literal values

- Results of a subquery

- Calculated from a function

Calculations can also be used in the WHERE and ORDER BY clause.

Aliases

All calculated values should be renamed with an alias. The word AS is optional, but is recommended to reduce confusion.

You can also alias noncalculated columns if the new name makes more sense.

Aliases are given in the SELECT clause, which is evaluated last before ORDER BY. For most DBMSs, this means that you can't use the alias in any other clause but the ORDER BY.

NULLs

A table may, of course, include NULLs in various places. As a rule, a NULL will wipe out any calculation, leaving NULL in its wake.

You can bypass NULLs with the coalesce() function which replaces NULL with an alternative value. You might also use a CASE ... END expression.

Casting Types

SQL works with three main data types:

- Numbers

- Dates and times

- Strings

You may need to change the data type. This is done with the cast() function.

When you cast within a major type, the effect is to change the precision or size of the type.

When you cast between major types, it is usually for compatibility. While casting *to* a string is usually possible and often automatic, casting *from* a string may not always succeed. Different DBMSs have different reactions to an unsuccessful cast.

Calculating with Numbers

You can perform basic arithmetic on all number types. Different DBMSs have various attitudes to working with integers.

SQL will include various functions to work with numbers, including

- Mathematical functions

- Approximation functions

There are also formatting functions which generate a formatted result as a string.

Calculating with Strings

Strings may be stored in various ways. Typically, a string uses ASCII or Unicode. String operations may or may not be case sensitive, depending on the collation of the database.

The basic simple operation with strings is concatenation. There is usually a simple operator to do this.

Other string operations involve string functions.

Calculating with Dates

Note that SQLite doesn't have a date data type. It does include some functions to convert strings or numbers to dates.

With dates, the following operations are common:

- Entering and storing a date and time

- Getting the current date and time

- Extracting part of the date or time

- Formatting a date

- Some simple arithmetic, such as the difference between dates and times, and modifying a date and time

The CASE Expression

The CASE expression allows you to choose from a number of alternative values. The case expression can be used to simplify values, as well as to group them.

There is a simple form of the CASE expression which can be used for discrete values.

CASE expressions can also be nested for more complex expressions.

Coming Up

Now that we've worked with table data, we can now start looking at analyzing it.

The next chapter will look at summarizing data with aggregate functions and grouping. We'll cover how data is aggregated in SQL, the basic aggregate functions, and summarizing into one or more groups.

We'll also look at combining aggregates at various levels, as well as some basic statistics on numerical data.

CHAPTER 5

Aggregating Data

Databases store data. That's obvious, but the data itself is pretty inert—you save it, you retrieve it, and you sometimes change it. That's OK for some things, but sometimes you want the data to work a little harder.

You can put the data to work when you start to summarize it. You can then see trends, see where it's going, or just get an overview of the data.

Aggregate functions are used to calculate summaries of data. They have three contexts:

- Summarize the whole table.

- Summarize in groups, using GROUP BY.

- Include summaries row by row. This is done with **window functions**, using the OVER clause.

You'll learn about window functions in Chapter 8. In this chapter, we look at how to calculate summaries, either wholly or in groups, using SQL's built-in aggregate functions.

The Basic Aggregate Functions

You've no doubt already had some experience with aggregate functions. The aggregate functions are basically statistical in nature and include

- count

  Count the number of values in a column, regardless of what the actual value is. As a special case, count(*) counts the number of rows in a table.

- sum and avg

© Mark Simon 2023
M. Simon, *Leveling Up with SQL*, https://doi.org/10.1007/978-1-4842-9685-1_5

Add or average the values in a column. Of course, you can only do this if the column is numeric.

- max and min

  Find the maximum or minimum value in a column. In case the interpretation is unclear, they find the first and last values you'd get if you used ORDER BY, except, of course, for the NULLS, which are always ignored.

- stddev, stddev_samp, stddev_pop (PostgreSQL, MySQL/MariaDB, Oracle) or stdev, stdevp (MSSQL)

  Find the standard deviation of the column values. This is either the population or sample standard deviation. Again, this only works with numeric columns.

  The actual name (and spelling) tends to vary slightly between DBMS. PostgreSQL treats stddev as a synonym for stddev_samp. MySQL/MariaDB treats it as a synonym for stddev_pop. Oracle treats it as a variation of stddev_samp.

There are various other aggregate functions, depending on the DBMS, but the preceding ones are fairly typical.

For example:

```
-- Book Data
SELECT
  -- Count Rows:
    count(*) AS nbooks,
  -- Count Values in a column:
    count(price) AS prices,
  -- Cheapest & Most Expensive
    min(price) AS cheapest, max(price) AS priciest
FROM books;
```

You get results like these:

nbooks	prices	cheapest	priciest
1201	1096	10	20

Or for numerical statistics:

```
--  Customer Data
    SELECT
    --  Count Rows:
        count(*) AS ncustomers,
    --  Count Values in a column:
        count(phone) AS phones,
    --  Height Statistics
        stddev_samp(height) AS sd    --  MSSQL: stdev(height)
      FROM customers;
```

You get results like these:

ncustomers	phones	sd
303	286	6.992

All of these functions are applicable to numbers, but only the following may be used for other data, such as strings and dates:

- count

- max and min

For example:

```
SELECT
--  Count Values in a column:
    count(dob) AS dobs,
--  Earliest & Latest
    min(dob) AS earliest, max(dob) AS latest
FROM customers;
```

gives you

Dobs	earliest	latest
239	1943-05-18	2003-01-27

We've been a little bit relaxed in the preceding descriptions. In particular

- A table may be a virtual table, such as a view, a join, or a common table expression.

- Any table with a WHERE clause will be filtered *before* the aggregates are applied.

- By value, we definitely don't include NULL. That's particularly apparent when you find that count() ignores NULLs and that avg() is divided by the non-NULL values.

- These functions apply to either the whole table or groups of rows.

NULL

Aggregate functions do not include NULLs. The only time this is not obvious is when using the sum function. However, it is significant to note that

- count(column) will only count the non-NULL values in the column, so you may get fewer than the total number of rows.

- avg(column) will also ignore the NULL values, so the average is divided only by the number of values, not necessarily the number of rows.

To put it another way, there is a world of difference between NULL on one hand and 0 or ' ' on the other.

We'll take advantage of this fact when we look at aggregate filters later.

Understanding Aggregates

Using aggregates sometimes runs into a few problems and seems to have a few quirky rules. It all makes more sense if you understand how aggregates really work.

When you aggregate data, the original data is effectively transformed into a new virtual table, with summaries for one or more groups.

For example, the query

```
SELECT
    count(*) AS rows,
    count(phone) AS phones
FROM customers;
```

can be regarded as

```
SELECT
    count(*) AS rows,
    count(phone) AS phones
FROM customers
GROUP BY () --  PostgreSQL, MSSQL, Oracle only
;
```

Note that the clause GROUP BY () doesn't work for all DBMSs, such as MariaDB/MySQL or SQLite. That doesn't matter, since the grouping is happening anyway.

The thing is, with or without the GROUP BY () clause, SQL will generate the virtual summary table as soon as it finds an aggregate function in the query.

In the preceding example, the data is summarized into a single virtual summary table of one row. In turn, this virtual table has grand totals for every column as in Figure 5-1.

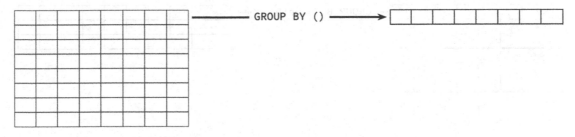

Figure 5-1. *Picturing GROUP BY ()*

This is why you can't include individual row data with an aggregate query. For example, this won't work:

```
SELECT
    id,                     --  oops
    count(*) AS rows,
    count(phone) AS phones
FROM customers;
```

You'll get an error message basically telling you that you can't use the id in the query.

Note that in MariaDB/MySQL in traditional mode, you can indeed run this statement successfully. However, the DBMS will pick the first id it can find, and that really has no meaningful value. It's mainly useful if you can be sure that all of the non-aggregate values are the same.

When you include a more meaningful GROUP BY clause, the result is similar, except that

- There is now one summary row for each group.

- There is also an additional column for each grouping column.

It looks something like Figure 5-2.

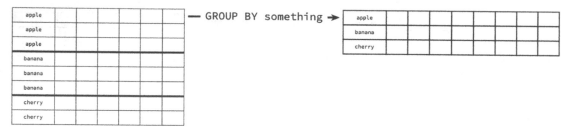

Figure 5-2. *GROUP BY Something*

For example:

```
SELECT
    town, state,            -- grouping columns
    count(phone) AS phones, -- summaries for each group:
    min(dob) AS oldest
FROM customerdetails
GROUP BY town, state;
```

You'll get something like this:

town	State	phones	oldest
[NULL]	[NULL]	24	1946-04-30
The Gap	QLD	5	1998-04-22
Lilydale	TAS	3	1945-08-31
Guildford	WA	3	1985-10-06
Kingston	VIC	2	1947-09-29
Reedy Creek	NSW	6	1960-12-30
~ 92 rows ~			

(You may get a group of NULLs either at the beginning or the end, because we haven't filtered out the NULL addresses.)

In the overall scheme of things, the (virtual) GROUP BY clause appears after the FROM and possibly WHERE clauses and is evaluated at that point:

```
SELECT ...
FROM ...
WHERE ...
GROUP BY ...
--  SELECT
ORDER BY ...
```

As usual, SELECT is evaluated last before ORDER BY, even though it is written first, as in Figure 5-3.

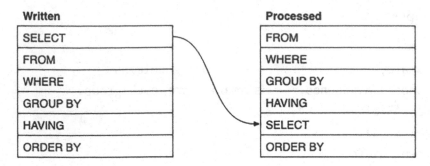

Figure 5-3. *Clause Order*

SQL neither knows nor cares about the actual meaning of the data, so there are no checks over whether you should apply these aggregate functions to particular columns.

Aggregating Some of the Values

In some cases, you may wish to aggregate only some of the values in a column. Here, we'll look at aggregating distinct values and then at filtering which values to aggregate.

Distinct Values

Most aggregate functions can be applied to distinct values, but it is probably statistically invalid. However, it can be meaningful if you *count* distinct values, such as in the following example:

```
SELECT
    count(state) AS addresses,
    count(DISTINCT state) AS states
FROM customerdetails;
```

This will count how many distinct states are in the customer details. That's not to say that you can't count the `state` column anyway, as it indicates the number of rows which have any address information at all:

Addresses	states
278	8

Be careful, though. It's possible that the column doesn't give the whole picture. For example, if you try

```
SELECT count(DISTINCT town) FROM customerdetails;
```

you'd get a result, but it might be open to misinterpretation. What you're getting is distinct *town names*, but many of these town names appear in more than one state. You shouldn't interpret this as meaning distinct *towns*.

As for the other aggregate functions, generally, it is meaningless to apply any other statistical calculation to only one of each sample.

Aggregate Filter

Normally, aggregate functions apply to the whole table or to the whole group. For example, count(*) will count all the rows in the table or group.

A relatively new feature allows you to apply an aggregate function to some of the rows. This can be applied multiple times in the query.

For example, the following will count all the customers in the customers table:

```
SELECT count(*) FROM customers;
```

Suppose you want to separate the customers into the younger and older customers. You might instinctively try something like this:

```
--  Don't bother trying this:
SELECT
      count(dob<'1980-01-01') AS older,
      count(dob>='1980-01-01') AS younger
  FROM customers;
```

If the preceding query doesn't produce an error, it is likely to be misinterpreted.

SQL does provide a working way of filtering what you want to filter. You can enter the following:

```
--  PostgreSQL:
SELECT
      count(*) FILTER (WHERE dob<'1980-01-01') AS older,
      count(*) FILTER (WHERE dob>='1980-01-01') AS younger
  FROM customers;
```

You get something like this:

Older	younger
133	106

Unfortunately, this is not well supported (currently, it is only supported in PostgreSQL). However, the following alternative will do the same:

```
SELECT
    count(CASE WHEN dob<'1980-01-01' THEN 1 END) AS old,
    count(CASE WHEN dob>='1980-01-01' THEN 1 END) AS young
FROM customers;
```

This uses the CASE expression to separate the dob values. They will either be 1 or NULL, and the count() function counts only the 1s.

You can also use this technique with other aggregate functions. For example:

```
-- New Standard
    SELECT
        sum(total),
        sum(total) FILTER (WHERE ordered <'...') AS older,
        sum(total) FILTER (WHERE ordered>='...') AS newer
    FROM sales;

-- Alternative
    SELECT
        sum(total),
        sum(CASE WHEN ordered<'...' THEN total END) AS older,
    SELECT
        sum(total),
        sum(CASE WHEN ordered<'...' THEN total END) AS older,
        sum(CASE WHEN ordered>='...' THEN total END) AS newer
    FROM sales;
```

Here, the value is either total or NULL, and sum() politely ignores the NULLs.

Sum	older	newer
342836.22	162045	164873.22

If you're interested in filtering for different categories, however, you might get more of what you want with grouping.

Grouping by Calculated Values

The preceding technique separates different groups horizontally—that is, each value is in the same row. You can also separate these derived groups vertically. This is achieved by using the GROUP BY clause.

You will be familiar with using GROUP BY with simple column values:

```
SELECT state, count(*)
FROM customerdetails
GROUP BY state;
```

You can also group by a derived value. For example, you can group your customers by their month of birth:

```
-- PostgreSQL, Oracle
   SELECT EXTRACT(month FROM dob) as monthnumber,
       count(*) AS howmany
   FROM customerdetails
   GROUP BY EXTRACT(month FROM dob)
   ORDER BY monthnumber;

-- MSSQL
   SELECT month(dob) AS monthnumber, count(*) AS howmany
   FROM customerdetails
   GROUP BY month(dob)
   ORDER BY monthnumber;

-- MySQL / MariaDB
   SELECT month(dob) AS monthnumber, count(*) AS howmany
   FROM customerdetails
```

```
    GROUP BY month(dob)
    ORDER BY monthnumber;

--  SQLite
    SELECT strftime('%m',dob) as monthnumber,
        count(*) AS howmany
    FROM customerdetails
    GROUP BY strftime('%m',dob)
    ORDER BY monthnumber;
```

In this example, the month number is called monthnumber, which is also used to sort the results.

Monthnumber	howmany
1	19
2	14
3	17
4	23
5	24
6	15
7	27
8	18
9	18
10	24
11	17
12	23
[NULL]	64

Note that the calculation appears twice, once in the SELECT clause and once in the GROUP BY clause. This is because the SELECT is evaluated after GROUP BY, so, alas, its alias is not yet available to GROUP BY.

This is not a real problem, as the SQL optimizer will happily reuse the calculation, so it's not really doing it twice.

Unfortunately, the month number isn't very friendly, so we could use the month name. However, inconveniently, the month name is in the wrong sort order, so we will need both:

```
-- Not SQLite

-- PostgreSQL, Oracle
   SELECT EXTRACT(month FROM dob) as monthnumber,
       to_char(dob,'Month') AS monthname,
       count(*) AS howmany
   FROM customerdetails
   GROUP BY EXTRACT(month FROM dob), to_char(dob,'Month')
   ORDER BY monthnumber;

-- MSSQL
   SELECT month(dob) AS monthnumber,
       datename(month,dob) AS monthname, count(*) AS howmany
   FROM customerdetails
   GROUP BY month(dob), datename(month,dob)
   ORDER BY monthnumber;

-- MySQL / MariaDB
   SELECT month(dob) AS monthnumber,
       monthname(dob) AS monthname, count(*) AS howmany
   FROM customerdetails
   GROUP BY month(dob), monthname(dob)
   ORDER BY monthnumber;
```

This looks better:

Monthnumber	monthname	howmany
1	January	19
2	February	14
3	March	17
4	April	23
5	May	24

(continued)

Monthnumber	monthname	howmany
6	June	15
7	July	27
8	August	18
9	September	18
10	October	24
11	November	17
12	December	23
[NULL]	[NULL]	64

As you see, you can't quite do this in SQLite since it doesn't have a function to get the month name.

Technically, grouping by both is redundant, since there is only one month name per month. However, we need both so that we can display one, but order by the other.

Although repeating the calculations is not a problem, it does make the query less readable and harder to maintain. We can take advantage of using a Common Table Expression:

```
WITH cte AS (
    ...
)
SELECT monthname, count(*)
FROM cte
GROUP BY monthnumber, monthname
ORDER BY monthnumber;
```

You can use GROUP BY with any calculated field, but note that

- Since simple calculations don't always result in something worth grouping, there is a limit on what you can do with them.

- As noted before, the calculation needs to be in both the SELECT clause and the GROUP BY clause, making the process tedious.

The second point earlier can be alleviated with the use of Common Table Expressions. The first point can be addressed by the use of CASE statements.

Grouping with CASE Statements

The basic GROUP BY presupposes that you already have values which can be grouped. Sometimes, such values can be derived, such as the month or day name.

More arbitrary groupings can be created using the CASE statement.

For example, suppose we want to count the younger and older customers. We can do this by using a CASE statement which distinguishes between them:

```
CASE
    WHEN dob<'1980-01-01' THEN 'older'
    WHEN dob IS NOT NULL then 'younger'
    -- ELSE NULL
END
```

Remember that some dobs may be NULL, so you need to filter them to get the younger ones. Remember, too, that the default ELSE is NULL, so we don't need to include it.

To count them, we could include this in the GROUP BY clause as follows:

```
SELECT count(*)
FROM customers
GROUP BY CASE
    WHEN dob<'1980-01-01' THEN 'older'
    WHEN dob IS NOT NULL then 'younger'
END;
```

This gives you something:

Count
64
133
106

but it's useless without some sort of labels. We can do this by repeating the calculation in the SELECT clause:

```
SELECT
    CASE
        WHEN dob<'1980-01-01' THEN 'older'
```

```
        WHEN dob IS NOT NULL then 'younger'
    END AS agegroup,
    count(*)
FROM customers
GROUP BY CASE
    WHEN dob<'1980-01-01' THEN 'older'
    WHEN dob IS NOT NULL then 'younger'
END;
```

This now works:

Agegroup	count
[NULL]	64
Older	133
Younger	106

but from the point of view of coding, it's worse than the calculated columns in the previous section, so this would definitely benefit from the use of a Common Table Expression:

```
WITH cte AS (
    SELECT
        *,
        CASE
            WHEN dob<'1980-01-01' THEN 'older'
            WHEN dob IS NOT NULL then 'younger'
        END AS agegroup FROM customers
)
SELECT agegroup,count(*)
FROM cte
GROUP BY agegroup;
```

This will now give you a more manageable result.

Revisiting the Delivery Status

Remember in a previous chapter we had created a delivery statistics using a nested CASE statement:

```
WITH salesdata AS (
--  PostgreSQL, MariaDB / MySQL, Oracle
    SELECT
        ordered, shipped, total,
        current_date - cast(ordered as date) AS ordered_age,
        shipped - cast(ordered as date) AS shipped_age
    FROM sales
--  MSSQL
    SELECT
        ordered, shipped, total,
        datediff(day,ordered,current_timestamp)
            AS ordered_age,
        datediff(day,ordered,shipped) AS shipped_age
    FROM sales
--  SQLite
    SELECT
        ordered, shipped, total,
        julianday('now')-julianday(ordered) AS ordered_age,
        julianday(shipped)-julianday(ordered) AS shipped_age
    FROM sales
)
SELECT
    ordered, shipped, total,
    CASE
        WHEN shipped IS NOT NULL THEN
            CASE
                WHEN shipped_age>14 THEN 'Shipped Late'
                ELSE 'Shipped'
            END
        ELSE
            CASE
```

```
            WHEN ordered_age<7 THEN 'Current'
            WHEN ordered_age<14 THEN 'Due'
            ELSE 'Overdue'
        END
    END AS status
FROM salesdata;
```

(Delete the unused SELECT statements, of course.) You get

Ordered	shipped	total	status
2022-05-15 21:12:07.988741	2022-05-23	28	Shipped
2022-05-16 03:03:16.065969	2022-05-24	34	Shipped
2022-05-16 10:09:13.674823	2022-05-22	58.5	Shipped
2022-05-16 15:02:43.285565	[NULL]	50	Overdue
2022-05-16 16:48:14.674202	2022-05-28	17.5	Shipped
[NULL]	[NULL]	13	Overdue
~ 5549 rows ~			

If you want to summarize this into status groups, you can again put the whole statement into a CTE and then summarize the CTE. You already have one CTE to precalculate the age, so we'll need another to hold the preceding results:

```
WITH
    salesdata AS (
        -- as above
    ),
    statuses AS (
        SELECT
            ordered, shipped, total,
            CASE
                WHEN shipped IS NOT NULL THEN
                    CASE
                        WHEN shipped_age>14
```

```
                       THEN 'Shipped Late'
                   ELSE 'Shipped'
               END
           ELSE
               CASE
                   WHEN ordered_age<7 THEN 'Current'
                   WHEN ordered_age<14 THEN 'Due'
                   ELSE 'Overdue'
               END
           END AS status
       FROM salesdata
    )
SELECT status, count(*) AS number
FROM statuses
GROUP BY status;
```

This will give you the summarized data:

Status	Number
Due	94
Current	78
Shipped	3808
Overdue	1273
Shipped Late	296

The next thing is to get the results in the right order.

Ordering by Arbitrary Strings

Of course, the real problem when it comes to sorting the results is that SQL has limited imagination and will only sort strings alphabetically. That only works well if the status values were also in alphabetical order, which they're not.

There are a few approaches you could take:

- You can include a number at the beginning of each string and then use ORDER BY. That's cheating and won't look right.

- You can have another table with the status values and a position number and then join this table to the main query. That's complicated, but may be useful in some cases.

- You can duplicate the CASE expression with numbers instead of the strings and ORDER BY that column instead. Unfortunately, there's no way to get two columns out of a single CASE expression. That's really messy.

- You ORDER BY the *position* of the string in a longer string.

We'll take the last approach earlier, since it's easy to implement and doesn't otherwise affect the results.

Most DBMSs include a function to find a substring in a larger string. It has various names and forms:

```
-- Postgresql
   POSITION(substring IN string)
-- MariaDB / MySQL & SQLite
   INSTR(substring,string)
-- Oracle
   INSTR(string,substring)
-- MSSQL
   CHARINDEX(substring,string)
```

In this case, we can find the position of the status string inside a longer string with the status values in order:

```
'Shipped,Shipped Late,Current,Due,Overdue'
```

The commas aren't necessary, but they make the string more readable. What's more important is that the status strings are in your preferred order, and the position function will return a lower value for strings it finds earlier. The rest is up to the ORDER BY clause.

We can order the preceding query using the positioning function like this:

```
WITH
    salesdata AS (
        -- as above
    ),
    statuses AS (
        -- as above
```

```
    )
SELECT status, count(*) AS number
FROM cte
GROUP BY status
--  Postgresql
    ORDER BY POSITION(status IN
        'Shipped,Shipped Late,Current,Due,Overdue')
--  MariaDB / MySQL & SQLite
    ORDER BY INSTR(status,
        'Shipped,Shipped Late,Current,Due,Overdue')
--  Oracle
    ORDER BY INSTR(status,
        'Shipped,Shipped Late,Current,Due,Overdue')
--  MSSQL
    ORDER BY CHARINDEX(status,
        'Shipped,Shipped Late,Current,Due,Overdue')
;
```

You'll now get the results in order:

Status	number
Shipped	3808
Shipped Late	296
Current	78
Due	94
Overdue	1273

You can use this technique for any nonalphabetical string order, such as days of the week or colors in the rainbow.

Group Concatenation

There is an additional function which can be used to aggregate string data. This function will concatenate strings with an optional delimiter.

183

This function has a few different names:

DBMS	Function
PostgreSQL	`string_agg(column, delimiter)`
SQL Server 2017+	`string_agg(column, delimiter)`
SQLite	`group_concat(column, delimiter)`
MySQL and MariaDB	`group_concat(column /* ORDER BY column */SEPARATOR delimiter)`
Oracle	`listagg(column, delimiter)`

For example, you can get a list of all the books for each author this way:

```
SELECT
    a.id, a.givenname, a.familyname,
    -- PostgreSQL, MSSQL
        string_agg(b.title, '; ') AS works
    -- SQLite
        -- group_concat(b.title, '; ') AS works
    -- Oracle
        -- listagg(b.title, '; ') AS works
    -- MariaDB / MySQL
        -- group_concat(b.title SEPARATOR '; ') AS works
FROM authors AS a LEFT JOIN books AS b ON a.id=b.authorid
GROUP BY a.id, a.givenname, a.familyname;
```

You'll get something like this:

id	givenname	familyname	works
146	Washington	Irving	Rip Van Wink ...; Tales of the ...; The ...
963	Richard	Marsh	The Beetle ...
390	Jean	Racine	Andromaque ...; Britannicus ...; Bérénice ...

(continued)

id	givenname	familyname	works
766	Evelyn	Everett-Green	True to the ...
296	Henri	Bergson	Matter and M ...; Laughter ...; Time and Fre ...
464	Ambrose	Bierce	An Occurrenc ...; The Monk and ...; Tales ...
~ 488 rows ~			

The works column has all of the book titles concatenated with a ; between them. Note that the GROUP BY clause uses the author id but includes the redundant author names to allow them to be selected.

Be careful, though. It's easy to get carried away with this function, and you'll see that the list of books can be very long, and the concatenated string can be very, very long.

Summarizing the Summary with Grouping Sets

Classically, using GROUP BY will give you totals for each combination of the groups. For example:

```
SELECT state, town, count(*)
FROM customerdetails
GROUP Y state, town;
```

will give you subtotals for each state/town combination.

Sometimes, you would like to include summaries of these subtotals, such as grand totals for each state and grand total overall.

Normally, you think of the word **total** as adding values and **subtotal** as a total of a subgroup. This would imply using the sum() function. In this discussion, we'll use the terminology more loosely and use the word for any aggregates, such as count(). Here, a subtotal would imply counting a subgroup.

In the preceding example, there are four possible totals that you could get:

- The count() of each state/town combination; this is what you normally get.

- The count() of each state group.

- The count() of each town group. In this example, it's not so useful, since some town names are duplicated across states, so you'd be combining values which shouldn't be. However, in other examples, this would be useful.

- The count() of the whole log—the grand total.

Apart from the last one, the others would all be considered subtotals at some level.

When we work with the example shortly, we'll aggregate by three columns, and there'll be eight combinations, so eight totals and subtotals we can calculate.

Modern SQL allows you to generate a result set which is a combination of totals and subtotals of table data and aggregate data. Depending on the DBMS, this might include a modification of the GROUP BY clause:

- GROUPING SETS allow you to specify which additional summaries to include. So, for example, you can decide which of the four possibilities earlier you want to include.

  This is supported by PostgreSQL, Microsoft SQL, and Oracle.

- ROLLUP is a simplified version of GROUPING SETS which produces *some* of the possible subtotals, treating the columns as a hierarchy. In the preceding example, you would get the state/town, state, and grand totals.

  This is supported by PostgreSQL, Microsoft SQL, Oracle, and MariaDB/MySQL.

- CUBE is also a specialized version of GROUPING SETS which produces *all* of the possible subtotals. In the preceding example, it's all four of the possible totals.

Here, we'll have a look at generating such a summary. However, rather than work with customers' addresses, we'll have a look at sales data.

Preparing Data for Summarizing

It's often the case that your original data isn't quite ready for summarizing. For example, the sales table includes the *time* of each order, but it's very hard to group that. For our sample, we'll work on summarizing the following:

- The month of the order

- The customer id

- The state the customer lives in

Note that all three columns are independent of each other, unlike the state and town in the original example. That means totaling any combination is meaningful.

To prepare the data, we can use the following query:

```
SELECT
    --  PostgreSQL, Oracle
        to_char(s.ordered,'YYYY-MM') AS ordered,
    --  MariaDB / MySQL
    --    date_format(s.ordered,'%Y-%m') AS ordered,
    --  MSSQL
    --    format(s.ordered,'yyyy-MM') AS ordered,
    --  SQLite
    --    strftime('%Y-%m',s.ordered) AS ordered,
    s.total, c.id, c.state
FROM sales AS s JOIN customerdetails AS c
    ON s.customerid=c.id
WHERE s.ordered IS NOT NULL;
```

You'll see something like this:

ordered	total	id	state
2022-05	28	28	NSW
2022-05	34	27	NSW
2022-05	58.5	1	WA
2022-05	50	26	VIC
2022-05	17.5	26	VIC
2022-07	15	105	VIC
~ 5295 rows ~			

No version of SQL has a straightforward way of extracting a year-month combination, so we use a formatting function. It returns a string, but that's fine for what we want to do with it.

187

When working with this, you could use this in a CTE, but it's not quite convenient, so we'll save it as a view instead:

```
DROP VIEW IF EXISTS salesdata;   --   Not Oracle
CREATE VIEW salesdata AS
SELECT
    --   PostgreSQL, Oracle
        to_char(s.ordered,'YYYY-MM') AS ordered,
    --   MariaDB / MySQL
        --   date_format(s.ordered,'%Y-%m') AS ordered,
    --   MSSQL
        --   format(s.ordered,'yyyy-MM') AS ordered,
    --   SQLite
        --   strftime('%Y-%m',s.ordered) AS ordered,
    s.total, c.id, c.state
FROM sales AS s JOIN customerdetails AS c
    ON s.customerid=c.id
WHERE s.ordered IS NOT NULL;
```

If you're using Microsoft SQL, remember to surround your CREATE VIEW statement between a pair of GOs:

```
-- MSSQL DROP VIEW IF EXISTS salesdata;
GO
    CREATE VIEW salesdata AS
    SELECT
        format(s.ordered,'yyyy-MM') AS ordered,
        s.total, c.id, c.state
    FROM sales AS s JOIN customerdetails AS c
        ON s.customerid=c.id
    WHERE s.ordered IS NOT NULL;
GO
```

To begin with, we'll generate the summaries separately and combine them with a UNION clause.

Combining Summaries with the UNION Clause

Most DBMSs have a simpler way of doing this, but you'll get a feeling for how all of this works doing it the (not very) long way.

To begin with, you can get summaries by state, customer ids, and dates using the following query:

```
-- All Group summaries
   SELECT state, id, ordered, count(*) AS nsales, sum(total) AS total
   FROM salesdata
   GROUP BY state,id,ordered
   ORDER BY state,id,ordered;
```

You'll get summaries for each state/customer id/ordered date combination.

state	id	ordered	nsales	total
ACT	85	2022-06	2	117
ACT	85	2022-07	2	104.5
ACT	85	2022-08	5	269.5
ACT	85	2022-09	3	253.5
ACT	85	2022-10	5	476
ACT	85	2022-11	3	179.5
~ 1802 rows ~				

The next step is to generate summaries for the state and customer ids:

```
-- state, ordered summaries
   SELECT
       state, id, NULL, count(*) AS nsales,
       sum(total) AS total
   FROM salesdata
   GROUP BY state, id
   ORDER BY state, id;
-- state summaries
   SELECT
```

```
      state, NULL, NULL, count(*) AS nsales,
      sum(total) AS total
   FROM salesdata
   GROUP BY state
   ORDER BY state;
```

You'll get something like the following:

state	id	?column?	nsales	total
ACT	85	[NULL]	37	2418.5
ACT	112	[NULL]	20	1272.5
ACT	147	[NULL]	32	2202
ACT	355	[NULL]	13	689.5
ACT	489	[NULL]	3	199
NSW	10	[NULL]	48	2931.5

~ 266 rows

and

state	?	?	nsales	total
ACT	[NULL]	[NULL]	105	6781.5
NSW	[NULL]	[NULL]	1668	102010.22
NT	[NULL]	[NULL]	103	6151
QLD	[NULL]	[NULL]	869	53331.5
SA	[NULL]	[NULL]	499	30977.5
TAS	[NULL]	[NULL]	456	28193
VIC	[NULL]	[NULL]	1273	79199.5
WA	[NULL]	[NULL]	322	20274

Don't worry about the missing column names, as we'll get them from the UNION.

The reason to include all those NULLs is to line up the columns when you combine them in a UNION.

Finally, get the grand total:

```
-- grand total
   SELECT
        NULL, NULL, NULL, count(*) AS nsales,
        sum(total) AS total
   FROM salesdata
   -- GROUP BY ()
   ;
```

This gives you a single row of grand totals:

?	?	?	nsales	total
[NULL]	[NULL]	[NULL]	5295	326918.22

Note that this includes the commented out GROUP BY () clause, just as a reminder that this is a grand total; of course, you don't need it.

The UNION clause can be used to combine the results of multiple SELECT statements. The only requirement is that they match in the number and types of columns.

```
-- All Group summaries
   SELECT
        state, id, ordered, count(*) AS nsales,
        sum(total) AS total
   FROM salesdata
   GROUP BY state,id,ordered
-- state, ordered summaries
   UNION
   SELECT state, id, NULL, count(*), sum(total)
   FROM salesdata
   GROUP BY state,id
-- state summaries
   UNION
   SELECT state, NULL, NULL, count(*), sum(total)
   FROM salesdata
   GROUP BY state
```

```
--  grand total
    UNION
    SELECT NULL, NULL, NULL, count(*), sum(total)
    FROM salesdata
--  Sort
    ORDER BY state,id,ordered;
```

You now have the results combined:

state	id	ordered	nsales	total
ACT	85	2022-06	2	117
ACT	85	2022-07	2	104.5
ACT	85	2022-08	5	269.5
ACT	85	2022-09	3	253.5
ACT	85	2022-10	5	476
ACT	85	2022-11	3	179.5
ACT	85	2022-12	2	84
ACT	85	2023-01	4	248.5
ACT	85	2023-03	3	209
ACT	85	2023-04	6	384
ACT	85	2023-05	2	93
ACT	85	[NULL]	37	2418.5
ACT	112	2022-07	2	72
ACT	112	2022-08	2	78
ACT	112	2022-09	1	49
ACT	112	2022-10	1	70.5

~ 2077 rows

Note that only the first query has aliases for the number of sales and the total; in a UNION, the column names for the first query apply to the whole result. You can alias the rest if it makes you feel better, but it won't make any difference.

When combining different levels of summaries, the higher-level summaries will have NULL instead of actual values. This is correct, but inconvenient:

- When sorted, NULL may appear at the beginning or the end of the list. The SQL standard is ambivalent on this, and different DBMSs have different opinions, while some give you a choice.

- In any case, NULL in the result set is unclear and unhelpful.

To resolve the sorting problem, we can add a contrived value to force a sorting order:

```
--   All Group summaries
SELECT
      state, id, ordered, count(*) AS nsales,
      sum(total) AS total,
      0 AS state_level, 0 AS id_level, 0 AS ordered_level
FROM salesdata
GROUP BY state,id,ordered
--   state, ordered summaries
UNION
SELECT
      state, id, NULL, count(*), sum(total),
      0, 0, 1
FROM salesdata
GROUP BY state,id
--   state summaries
UNION
SELECT
      state, NULL, NULL, count(*), sum(total),
      0, 1, 1
FROM salesdata
GROUP BY state
--   grand total
UNION
SELECT
      NULL, NULL, NULL, count(*), sum(total),
      1, 1, 1
```

```
    FROM salesdata
--  Sort
    ORDER BY state_level, state, id_level, id,
        ordered_level, ordered;
```

To get the results in the right order, we have introduced two values, state_level and town_level, so that we can push the totals below the other values.

state	id	ordered	nsales	total	state_level	id_level	ordered_level
ACT	85	2022-06	2	117	0	0	0
ACT	85	2022-07	2	104.5	0	0	0
ACT	85	2022-08	5	269.5	0	0	0
ACT	85	2022-09	3	253.5	0	0	0
ACT	85	2022-10	5	476	0	0	0
ACT	85	2022-11	3	179.5	0	0	0
ACT	85	2022-12	2	84	0	0	0
ACT	85	2023-01	4	248.5	0	0	0
ACT	85	2023-03	3	209	0	0	0
ACT	85	2023-04	6	384	0	0	0
ACT	85	2023-05	2	93	0	0	0
ACT	85	[NULL]	37	2418.5	0	0	1
ACT	112	2022-07	2	72	0	0	0
ACT	112	2022-08	2	78	0	0	0
ACT	112	2022-09	1	49	0	0	0
ACT	112	2022-10	1	70.5	0	0	0

~ 2077 rows ~

To eliminate the sorting columns from the result set, you can turn this into a Common Table Expression:

```
WITH cte AS (
    -- UNION query above
)
SELECT state, id, ordered, nsales, total
FROM cte
ORDER BY state_level,state,id_level,id,ordered_level,ordered;
```

This isn't so much work to get the results, but there may be a simpler method.

Using GROUPING SETS, CUBE, and ROLLUP

Grouping sets offer a simpler alternative to the UNION above. Their syntax is not immediately obvious, but it follows a similar pattern to the GROUP BY clauses.

Grouping sets are not fully supported by all DBMSs. Most support a simpler version, but SQLite doesn't support them at all. If you have SQLite, the best you can do is what we did earlier with the UNION.

For the most part, what you probably want is the ROLLUP described later.

GROUPING SETS and CUBE (PostgreSQL, MSSQL, and Oracle)

The most general-purpose technique uses the GROUPING SET clause. In this clause, you specify which combinations of columns you want to include in the summary.

The syntax is

```
SELECT columns
FROM table
GROUP BY GROUPING SETS ((set),(set));
```

Recall that the previous example had SELECT statements, grouped by state, customer id, ordered date, and a grand total. This can be generated as follows:

```
SELECT state,town,count(*)
FROM customers
GROUP BY GROUPING SETS ((state,id,ordered),(state,id),(state),());
```

Here, the set () indicates the grand summary.

Many other combinations are available (such as (state,ordered)). If you really wanted them all, you can use CUBE:

```
SELECT state, id, ordered, count(*), sum(total)
FROM salesdata
GROUP BY CUBE (state,id,ordered)
```

The CUBE variation works best when you don't have too many grouping columns and when they're all unrelated to each other. Remember three columns would give you eight possible combinations. You can calculate the number of possibilities as 2^n, where n is the number of columns. In this case, it's $2^3 = 8$. If you had even four columns, you would have 16 possible totals and subtotals, which might start to get overwhelming.

USING ROLLUP (PostgreSQL, MSSQL, Oracle, and MariaDB/MySQL)

A simpler, more readable (but slightly less flexible) alternative is to use ROLLUP. This takes two forms:

```
--  ROLLUP(...)        -   PostgreSQL, MSSQL, Oracle
    SELECT state, id, ordered, count(*), sum(total)
    FROM salesdata
    GROUP BY ROLLUP (state, id, ordered);

--  ... WITH ROLLUP    -   MariaDB / MySQL, MSSQL
    SELECT state, id, ordered, count(*), sum(total)
    FROM salesdata
    GROUP BY state, id, ordered WITH ROLLUP;
```

Both forms will give you the same result. Note that MSSQL gives you the choice to use either form.

ROLLUP makes an important assumption that the columns form some sort of hierarchy. In the case of the customer state and the customer id, that's obvious. Whether you consider the ordered date as the end of the hierarchy is up to you.

You can see the hierarchy in the results and in the fact that this matches the GROUPING SETS example earlier. You will get results for

1. (state, id, ordered) combinations

2. (state, id) combinations

3. (state) values

4. () – grand totals

Clearly, using ROLLUP is a much simpler way to get these results, and you probably won't miss the flexibility of GROUPING SETS very much.

Sorting the Results

In some DBMSs, you may see the results sorted automatically in the same order as the UNION version earlier. In some, you will need to sort it yourself.

This again has the two problems mentioned with the UNION version earlier. To address the second problem, that of sort order, we can use the grouping(column) function which indicates the level of a column being summarized. The value is 1 to indicate that this is a summary and 0 to indicate that it's not. This will have the same effect as the preceding contrived level columns:

```
-- ROLLUP(...):      PostgreSQL, MSSQL, Oracle
   SELECT state, id, ordered, count(*), sum(total)
   FROM salesdata
   GROUP BY ROLLUP (state,id,ordered)
   ORDER BY grouping(state), state, grouping(id), id,
       grouping(ordered), ordered;

-- ... WITH ROLLUP: MySQL, MSSQL
-- (MariaDB doesn't support grouping())
   SELECT state, id, ordered, count(*), sum(total)
   FROM salesdata
   GROUP BY state,id,ordered WITH ROLLUP
   ORDER BY grouping(state), state, grouping(id), id,
       grouping(ordered), ordered;`
```

You'll get a better sorted table.

Note that while MySQL does support grouping(), MariaDB *doesn't* support the grouping() function!

In PostgreSQL and MySQL, you can use grouping() with multiple columns. This will give you a combined level which is a binary combination of the 1s and 0s. In MSSQL and Oracle, you would use the grouping_id() function for that.

To solve the first problem, that of meaningless NULL markers, we have to be more creative with the SELECT clause. In this case, we can use coalesce to pick up the NULL and supply an alternative value:

```
--  PostgreSQL, MSSQL;
    SELECT
        coalesce(state,'National Total') AS state,
        coalesce(cast(id as varchar),state||' Total') AS id,
        coalesce(ordered,'Total for '||cast(id as varchar))
            AS ordered,
        count(*), sum(total)
    FROM salesdata
    GROUP BY ROLLUP (state,id,ordered)
    ORDER BY grouping(state), state,
        grouping(id), id, grouping(ordered), ordered;

--  MySQL, MSSQL (MariaDB doesn't support grouping())
    SELECT
        coalesce(state,'National Total') AS state,
        coalesce(cast(id as varchar),state||' Total') AS id,
        coalesce(ordered,'Total for '||cast(id as varchar))
            AS ordered,
        count(*), sum(total)
    FROM salesdata
    GROUP BY state,id,ordered WITH ROLLUP
    ORDER BY grouping(state), state,
        grouping(id), id, grouping(ordered), ordered;

--   NOT Oracle
```

This will give you something meaningful for the summary rows.

Renaming Values in Oracle

You'll notice that this won't work for Oracle. Here, Oracle proves to be unhelpful since
(a) NULL strings are equivalent to an empty string (' ') so they won't coalesce, (b) the
selected columns are no longer considered to match the GROUP BY clause, and (c) you'll
also have problems sorting with the grouping() function when you've made the other
changes.

The most straightforward way of getting this to work is to use a CASE expression for
the columns:

```
SELECT
    coalesce(state,'National Total') AS state,
    grouping(state) AS statelevel,
    CASE
        WHEN state IS NULL THEN NULL
        WHEN id IS NULL THEN 'Total for '||state
        ELSE cast(id AS varchar(3))
    END AS id,
    grouping(id) AS idlevel,
    CASE
        WHEN id IS NULL THEN NULL
        WHEN ordered IS NULL THEN
            'Total for '||cast(id as varchar(3))
        ELSE ordered
    END AS ordered,
    grouping(ordered) AS orderedlevel,
    count(*) AS count, sum(total) AS sum
FROM salesdata
GROUP BY ROLLUP (state,id,ordered)
ORDER BY statelevel, state, idlevel, id, orderedlevel, ordered
;
```

It will look something like this:

state	statelevel	id	idlevel	ordered	orderedlevel	count	sum
ACT	0	112	0	2022-07	0	2	72
ACT	0	112	0	2022-08	0	2	78
ACT	0	112	0	2022-09	0	1	49
ACT	0	112	0	2022-10	0	1	70.5
ACT	0	112	0	2022-11	0	1	94
ACT	0	112	0	2022-12	0	4	224
ACT	0	112	0	2023-01	0	1	48.5
ACT	0	112	0	2023-02	0	5	320
ACT	0	112	0	2023-03	0	2	191.5
ACT	0	112	0	2023-05	0	1	125
ACT	0	112	0	Total for 112	1	20	1272.5
ACT	0	147	0	2022-08	0	8	392
ACT	0	147	0	2022-09	0	2	199.5
ACT	0	147	0	2022-10	0	2	162
ACT	0	147	0	2022-11	0	3	228.5
ACT	0	147	0	2022-12	0	3	251

~ 2077 rows ~

Here, the grouping() function is used in the SELECT clause and then used for sorting. The id and ordered columns are calculated with a CASE ... END expression to get around the problem of the NULL strings.

Of course, now you have those three extra columns used for sorting. To hide them, you can use a CTE:

```
WITH cte AS (
    -- SELECT statement as above
    --  don't bother with the ORDER BY clause
)
```

```
SELECT state, id, ordered, count, sum
FROM cte
ORDER BY statelevel, state, idlevel, id, orderedlevel, ordered
;
```

Incidentally, the previous statement included an ORDER BY clause. You can include it in the CTE (you can't in MSSQL), but it's unnecessary as we're sorting it anyway in the main query, so you should leave it out.

Histograms, Mean, Mode, and Median

You may have learned some simple statistics at school. Here's a reminder of some of the basic concepts:

- A **frequency table** is a table of values and how often they appear.

- A **mean**, or more technically an **arithmetic mean**, is what we casually call the **average**: it's the total divided by the number of values.

- The **mode** is the value which appears the most often.

- The **median** is the middle of the values (if they're all placed in order).

You can use the frequency table to generate a **histogram**, which is what spreadsheet programs call a bar chart. For example, you can generate a frequency table and histogram of the number of customers per height (in centimetres). It looks like Figure 5-4.

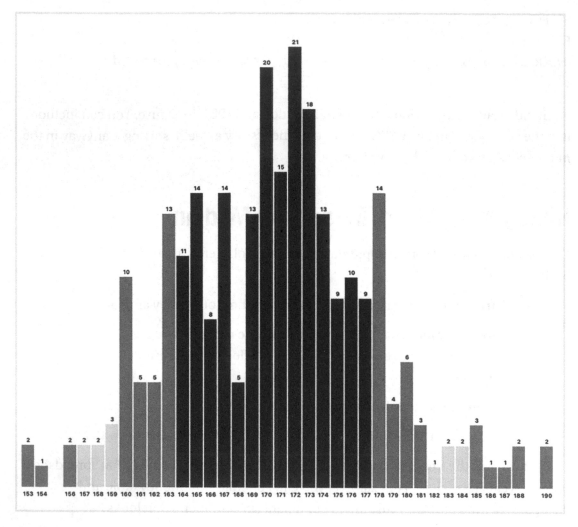

Figure 5-4. *A Histogram*

If your values are based on a number of different factors, there is a tendency for them to be distributed along the well-known "bell curve." Most values occur around the middle, and the further you are from the middle, the fewer times the value occurs. This is more technically referred to as the **normal distribution**.

Your height is dependent on a number of factors, some of which include genetics, diet, and other lifestyle factors. As a result, customer heights tend to follow the normal distribution as you see very roughly in the figure above. Of course, the tendency becomes stronger if we have a larger collection of data: if you have only a few hundred samples, then the data won't be such a tight fit.

For the purpose of this discussion, we'll focus on customer heights, as they tend to be easy to analyze this way.

Although the sample data was randomized, it was generated to follow the normal distribution as well as might be expected in the small sample.

For adults in Australia, the mean height is about 168.7 cm. Actually, there are two mean heights, one for female and one for male adults, but between them the average is 168.7 cm. The standard deviation is 7 cm. You can get more information at `https://en.wikipedia.org/wiki/Average_human_height_by_ country`.

Calculating the Mean

In statistics and mathematics, the word "average" is a little vague, as there are many measurements which can describe an average. The one we're interested in here is called arithmetic mean. You already have the mean as a built-in aggregate function `avg()`:

```
SELECT avg(height) AS mean FROM customers;
```

You'll get something like this:

mean
170.844

If you want, you can calculate the mean using `sum(height)/count(height)`, but there's no point other than to show they're the same.

This figure is fairly reasonable. The mean height for adults is about 168.7 cm.

Generating a Frequency Table

A frequency table lists how many times a value appears in the table. If that sounds suspiciously like something you've done before, you're right. That's just using a GROUP BY with `count(*)`.

First, however, we need to make the data more groupable. If the data is too fine, you won't get many duplicates—the irony is that there can be too much detail to summarize.

Here, the heights are measured to 0.1 of a centimeter. We'll prepare it by rounding it off to whole centimeters:

```
SELECT floor(height+0.5) AS height
FROM customers
WHERE height IS NOT NULL;
```

We've also filtered out the missing heights to give you

Height
169
171
153
176
156
176
~ 267 rows ~

We could possibly have used a round() function to do the rounding off, but some DBMSs prefer to round to the nearest even number, so the preceding method will do more reliably.

Putting the data into a CTE, we can then use a simple GROUP BY query:

```
WITH heights AS (
    SELECT floor(height+0.5) AS height
    FROM customers
    WHERE height IS NOT NULL
)
SELECT height, count(*) AS frequency
FROM heights
GROUP BY height
ORDER BY height;
```

This will get you

Height	frequency
153	1
154	3
156	1
157	3
158	1
159	2
~ 36 rows ~	

Note that there may be some missing values. That's natural, especially in a relatively small sample such as we have. However, with these gaps it's not quite ready for a histogram. Later, when we have a closer look at recursive common table expressions, we'll see how to fill in the gaps.

Calculating the Mode

The mode is the value which occurs most often. Actually, there may be more than one mode which tie for first.

To get the mode(s), you first need to get the frequencies and then find the maximum frequency. In principle, you're looking for max(count(*)), but you can't nest aggregate functions like that. That means taking an extra step, which, in turn, means more CTEs.

Starting with the previous frequency table, we can put the results in another CTE:

```
WITH
    heights AS (
        SELECT floor(height+0.5) AS height
        FROM customers
        WHERE height IS NOT NULL
    ),      -- don't forget to add a comma here
    frequency_table AS (
        SELECT height, count(*) AS frequency
```

```
        FROM heights
        GROUP BY height
    )
...
```

You'll also need the maximum frequency, which has to be calculated as a separate step:

```
WITH
    heights AS (
        ...
    ),
    frequency_table AS (
        ...
    ),      --  don't forget to add a comma here
    limits AS (
        SELECT max(frequency) AS max FROM frequency_table
    )
...
```

Finally, you can cross join the frequency table to the limits CTE to find the mode(s):

```
WITH
    heights AS (
        ...
    ),
    frequency_table AS (
        ...
    ),
    limits AS (
        ...
    )
SELECT height, frequency
FROM frequency_table,limits
WHERE frequency_table.frequency=limits.max
ORDER BY height;
```

You'll get something like this:

Height	frequency
172	22

In a perfect set of normal data, the mode should match the mean exactly. In real life, it should be close.

Calculating the Median

The median is the middle value. What that means is that half of the values should be below the median and half should be above.

To find the median involves putting all of the values in order and finding the midpoint. You can do this if you like, but that involves some skills we haven't developed yet, in particular getting the row number. We'll look at that later in the chapter on window functions.

Fortunately, modern SQL includes a function called `percentile_cont`. Unfortunately, not all DBMSs use it the same way, and SQLite doesn't support it at all.

The `percentile_cont()` function finds the value by its percentile. A percentile is a grouping of 100 groups. The 50th percentile would be in the middle.

To find the median in PostgreSQL:

```
-- PostgreSQL: aggregate function only
SELECT percentile_cont(0.5)
     WITHIN GROUP (ORDER BY height)
FROM customers
WHERE height IS NOT NULL;
```

This should give something like

percentile_cont
171.2

As noted in the comment, in PostgreSQL `percentile_cont()` is an aggregate function only, which is exactly what we want anyway.

What's the alternative to an aggregate function? It's one of those window functions which we'll be looking at later. A window function is like an aggregate function, except that it's calculated for every row, not just as a summary.

With the window function version, we can use

```
-- MySQL / MariaDB, MSSQL, Oracle: window function only
   SELECT percentile_cont(0.5)
       WITHIN GROUP (ORDER BY height) OVER()
   FROM customers
   WHERE height IS NOT NULL;
```

The problem is that you'll get the same value for multiple rows. To finish the job, you can use DISTINCT:

```
-- MySQL / MariaDB, MSSQL, Oracle: window function only
   SELECT DISTINCT percentile_cont(0.5)
       WITHIN GROUP (ORDER BY height) OVER()
   FROM customers
   WHERE height IS NOT NULL;
```

This will now give you the median.

The Standard Deviation

The standard deviation is a measure of how varied the data is. For the normal distribution, about 68% of the results should fall within one standard deviation of the mean.

A lower than expected standard deviation would indicate that the results are fairly close together, while a higher than standard deviation would indicate that the results are more random than expected. In either case, you would suspect that the data is biased—either by how the data was selected or suggesting that some of it was made up.

For example, if the heights were normally distributed, you'd find that one standard deviation amounts to 7 cm and that 68% of the customers should be within 7 cm on either side of the mean.

To calculate the standard deviation, SQL has a simple aggregate function, which you have already seen earlier.

You might wonder why there are two. If you know that you have all the values, you could use `stddev_pop(height)` (or `stdevp(height)` for MSSQL). However, we don't, so we can regard what we do have as a sample. For that, we use `stddev_samp(height)` (or `stdev(height)` for MSSQL):

```
SELECT
    stddev_pop(height) AS sd
    --  stdevp(height) AS sd          --  MSSQL
FROM customers;
```

You should get something close to 7:

Sd
6.979

Remember that the standard deviation only has meaning when you believe that the underlying data follows a normal distribution.

Summary

In this chapter, we had a look at aggregating sets of data.

Basic Aggregate Functions

The basic aggregate functions are

- `count()`, which counts the number of values in a column or rows in a table.

- `sum` and `avg`, which add or average numbers in a column.

- `max` and `min`, which find the lowest and highest values of any type of column. In effect, they find the first and last values when sorted by that column.

- `stddev`, `stddev_samp`, `stddev_pop` (PostgreSQL, MySQL/MariaDB, Oracle) `stdev`, `stdevp` (MSSQL). This calculates a population or standard deviation of a column of numbers; we assume that the data is normally distributed.

The sum, avg, and standard deviation functions can only be applied to numeric data.

NULLs

Aggregate functions all skip NULLs. This is particularly important when counting values, but also when calculating averages.

The fact that NULLs are skipped can also be used when calculating selective aggregates.

The Aggregating Process

The data you aggregate is not necessarily in a saved table. You can also aggregate virtual tables generated by joins, common table expressions, and views.

When aggregating data, the data may first be filtered by a WHERE clause. The aggregate data is based on what's left after filtering.

Performing aggregates effectively transforms the original data into a virtual table of one or more summary rows. This no longer includes the original raw data. You can only select what is in the virtual summary table. As a result, you can't mix aggregates with raw data in a single SELECT clause.

If you need both original and summary data, you may need to join the original table with a CTE which has the summary.

Aggregate Filters

It's possible to filter what data is used for a single aggregate function.

There is a standard FILTER (WHERE ...) clause which allows you to filter a column. However, it's not (yet) widely supported.

The common way to filter data is to use the CASE ... END expression on what you're aggregating. Set a value of, say, 1 for the values you want, allow the rest to default to NULL, and let the aggregate functions ignore them for the rest.

You can also aggregate on DISTINCT values. This makes the most sense when you are counting.

GROUP BY

The GROUP BY clause can be used to generate a virtual table of group summaries.

In some DBMSs, you can use GROUP BY () to generate a grand summary. This is the default without the GROUP BY () clause and is automatically done whenever SQL sees an aggregate function. It's never truly needed.

You can group by basic values, but also by calculated values.

Grouping by calculated values can get complicated, since the SELECT and ORDER BY clauses can only use what's in the GROUP BY. Because of the clause order, you may find yourself repeating the same calculations in various clauses.

Since the SELECT clause is only evaluated near the end, and selecting and ordering can only be done on what's in the GROUP BY clause, you may find the following techniques helpful:

- Using redundant groups to select one thing and sort by another

- Putting aggregate queries in a CTE and joining that with other tables to get the rest of the results

When grouping by a column, your results may not be in the correct order. Since the group names are all strings, sorting on the group name will only put them in alphabetical order, which isn't always suitable. However, you can also sort them by their position in another string, which can be in any order you like.

Mixing Subtotals

By and large, aggregate queries produce simple aggregates on one level. Sometimes, you need to combine them with various levels of subtotals.

You can generate subtotals in separate queries and combine them with UNION. You might need some extra work to get the results sorted in your preferred order.

Most DBMSs include subtotaling operations to create the combined result automatically. They may include GROUPING SETS, ROLLUP, or CUBE. Most include the ROLLUP which is the most common variation. There are additional grouping functions to assist with sorting and labeling.

Statistics

In general, aggregate functions are basically statistical in nature. Although SQL is not as powerful as dedicated statistical software, you can use aggregates and grouping to generate some of the basic statistics.

Coming Up

In some cases, we have used a query in a Common Table Expression to prepare data. However, in one case we created a view instead, so that we could reuse the query.

In the next chapter, we'll have a closer look at creating and using views to improve our workflow.

## CHAPTER 6

Using Views and Friends

This isn't the first time you've looked at views in this book, and it won't be the last time.

This chapter consolidates what you may have already picked up about views and related concepts and gives a few ideas about working with them in general.

You can spend the rest of your life writing SQL statements, and the job would get done. However, you might get to the point where writing the same thing over and over again loses its charm, and so you'll want to find ways of reusing previous queries.

First, let's have a look at what we mean by tables and what happens when you use the `SELECT` statement.

SQL databases store data in tables. Actually, they don't—each table is really stored in some other structure such as a binary tree, which is more efficient. However, by the time you see it, it will be presented as a table, and that's what it's called in the database.

A table is made up of rows and columns. For our purpose, the table doesn't have to be a permanent table, and there are operations which generate table structures without necessarily being permanently stored. We'll refer to them as **virtual tables**.

Here is a list of operations which generate (virtual) tables, in increasing order of longevity:

- The result of a `SELECT` statement is a virtual table.

- A **join** is the combination of two or more (virtual) tables to produce an expanded virtual table.

- A **Common Table Expression** generates a virtual table which you can use later in the query. A table subquery amounts to the same thing.

213

© Mark Simon 2023
M. Simon, *Leveling Up with SQL*, https://doi.org/10.1007/978-1-4842-9685-1_6

- A **view** is a saved SELECT query, which will regenerate the virtual table on call.

  A **materialized view** is a view which stores its results so it doesn't always have to regenerate everything.

  Some DBMSs support **Table-Valued Functions** which are functions which generate virtual tables.

- A **temporary table** is like a real table. It may or may not be stored on the disk. It will self-destruct when the session is finished.

The thing about tables and virtual tables is that they can all be used in the FROM clause.

You will already know about using joins. You have also used Common Table Expressions, but we'll discuss them in more depth in the following chapters.

In this chapter, we'll look at the rest and how we can improve our workflow with them.

Working with Views

You have already created one or two views so far, so much of this is familiar territory.

A view is a SELECT statement saved in the database. When the view is saved, the DBMS often also saves the execution plan, so less time is wasted planning the same thing next time.

Views can be used whenever you might have read from a table. You can think of a view as a virtual table—something which behaves as a table.

To read from a view, you treat it as a table:

```
SELECT columns FROM view;
```

Note the syntax is exactly the same as for tables. From the perspective of a SELECT statement, there is no distinction between selecting from a view and from a table.

One important consequence of this is that you cannot have views with the same names as tables—views and tables share the name space.

This doesn't mean that there are no differences. The DBMS stores views as separate types of objects and manages them differently. However, once created, you can treat a view like a table.

Views can be an important part of your workflow. For example:

- Saving a complex query to be used simply

- Exposing complex queries to external applications

- Creating an interface between existing data and an application

- As a substitute for some queries

Creating a view requires permissions which you may not already have as a database user.

Do whatever is required to get these permissions—badger, bribe, blackmail as needed.

Views have several benefits:

- You can hide complex processing with a view which just gives a virtual table of the results.

- You can restrict access to data in tables by creating a view with only certain rows and columns.

- You can hide details in aggregate queries.

There are some limitations as well. The main limitation is that a view is inflexible. You can't, for example, vary the values that a view might use to calculate results. You'll see a possible solution to that when we discuss Table Valued Functions later.

Other limitations vary between DBMSs. Some DBMSs support temporary views. MSSQL doesn't allow an ORDER BY in a view without additional trickery. Some DBMSs support views on temporary tables, while others don't.

For the most part, however, views will simplify your workflow and allow you to move on to more complex tasks.

Creating a View

A view starts off as a simple SELECT statement. For example, we can start developing a pricelist view which will comprise some information about books, their authors, and the price, including tax:

```
/* Notes
    =====================================================
    MSSQL:  Use + for concatenation
    Oracle: No AS for tables:
              FROM books b JOIN authors a ON ...
    ===================================================== */
SELECT
    b.id, b.title, b.published,
    coalesce(a.givenname||' ','')
        || coalesce(othernames||' ','')
        || a.familyname AS author,
    b.price, b.price*0.1 AS tax, b.price*1.1 AS inc
FROM books AS b LEFT JOIN authors AS a ON b.authorid=a.id
WHERE b.price IS NOT NULL;
```

This should give you something like this:

id	title	pub	author	price	tax	inc
2078	The Duel	1811	Heinrich von Kleis ...	12.50	1.25	13.75
503	Uncle Silas	1864	J. Sheridan Le Fan ...	17.00	1.70	18.70
2007	North and South	1854	Elizabeth Gaskell	17.50	1.75	19.25
702	Jane Eyre	1847	Charlotte Brontë	17.50	1.75	19.25
1530	Robin Hood, The Pr ...	1862	Alexandre Dumas	12.50	1.25	13.75
1759	La Curée	1872	Émile Zola	16.00	1.60	17.60

~ 1096 rows ~

As it's a price list, we've filtered out the NULL prices.

There are, however, certain conditions:

- All columns in a view must have a name, including calculated columns.

  You can't have anonymous columns as you might in a simple SELECT statement.

- Column names must be unique.

  This is especially important with joined tables, where column names may be duplicated.

In other words, the virtual table must conform to the rules of a real table.

To create a view, prepend the SELECT statement with a CREATE VIEW ... AS clause:

```
/* Notes
    =====================================================
    MSSQL:  Use + for concatenation
    MSSQL:  Surround the CREATE VIEW with GO
    Oracle: No AS for tables:
            FROM books b JOIN authors a ON ...
    ================================================= */
-- GO
    CREATE VIEW aupricelist AS
    SELECT
        b.id, b.title, b.published,
        coalesce(a.givenname||' ','')
            || coalesce(othernames||' ','')
            || a.familyname AS author,
        b.price, b.price*0.1 AS tax, b.price*1.1 AS inc
    FROM books AS b JOIN authors AS a ON b.authorid=a.id;
-- GO
```

That will now be saved into your database.

We've called the price list `aupricelist` because the tax is set to 10%, which is the rate in Australia. Feel free to use any tax rate and name that you like.

> In Microsoft SQL Server only, you need to separate the `CREATE` view statement from the rest of the script. You do this by putting `GO` before and after, which marks off what they call a batch. The `GO` must be on a line by itself.

You can now read the appropriate results like a table:

```
SELECT * FROM aupricelist;
```

This will give you the same results as before.

You can filter the view as with a table:

```
SELECT *
FROM aupricelist
WHERE published BETWEEN 1700 AND 1799;
```

This gives you

id	title	pub	author	price	tax	inc
1608	The Autobiography ...	1791	Benjamin Franklin	18.50	1.85	20.35
2303	The Metaphysics of ...	1797	Immanuel Kant	12.00	1.20	13.20
1305	An Essay on Critic ...	1711	Alexander Pope	11.00	1.10	12.10
1963	A Treatise of Huma ...	1740	David Hume	18.50	1.85	20.35
1196	Equiano's Travels: ...	1789	Olaudah Equiano	12.50	1.25	13.75
1255	Discourse on the O ...	1755	Jean-Jacques Rouss ...	19.00	1.90	20.90

~ 166 rows ~

You can also sort the results:

```
SELECT * FROM aupricelist ORDER BY title;
```

This gives you

id	title	pub	author	price	tax	inc
541	120 Days of Sodom	1904	Marquis de Sade	12.50	1.25	13.75
729	A Cartomante e Out ...	1884	Machado de Assis	16.00	1.60	17.60
2092	A Chaste Maid in C ...	1613	Thomas Middleton	15.00	1.50	16.50
1437	A Child's Garden o ...	1885	Robert Louis Steve ...	11.00	1.10	12.10
454	A Christmas Carol	1843	Charles Dickens	13.50	1.35	14.85
1094	A Confession	1882	Leo Tolstoy	17.50	1.75	19.25

~ 1096 rows ~

With the exception of MSSQL, you could have included the ORDER BY clause in the view itself. Although it's convenient, it's probably not a good idea: you forcing the DBMS to sort the result whether you need it or not, and you may end up sorting it again in a different order afterward.

Using ORDER BY in MSSQL

As noted earlier, SQL Server precludes ORDER BY in a view without additional trickery. If you *really* want to have a presorted view, you can do the following:

```
CREATE VIEW something AS
SELECT columns
FROM table
ORDER BY columns OFFSET 0 ROWS;
```

Among other things, this will allow you to create an ordered view without the need to include extra columns just for sorting.

However, you need to be aware that an ordered view does place an extra burden on the database, so it should only be used when needed.

Tips for Working with View

Views are generally a good thing, but there are a few things to help you use them more reliably.

Some of the fine points will vary between DBMSs, and the DBMS will do its best to work as efficiently as possible. Nevertheless, it's a good idea to keep these ideas in mind.

Don't Cascade Views Too Much

You can build a view using other views. That can help simplify your workflow, but you need to be careful:

- Changing the underlying view may make a mess of the view. Some DBMSs won't even allow you to change a view if another view depends on it.

- SQL tries to optimize your queries, but if your views are too deeply nested, it may not be able to optimize well.

- One of the views may have more than you need for the new view, so you're wasting processing time generating what you don't need.

That doesn't mean you shouldn't build on existing views, just that you should do it judiciously.

Don't Use SELECT *

There's a good chance that SELECT * doesn't give you quite what you want anyway, but, even if it does, you should probably list all of the columns specifically. Some DBMSs will even convert SELECT * to a column list.

Reasons why SELECT * may not be a good idea include

- You may want your own column order rather than the default one.

- The underlying table may undergo a change in structure, so your view may not end up being what you originally expected.

Avoid Using ORDER BY

Most DBMSs allow you to include an ORDER BY clause, and, even in MSSQL, you can force it to. However, it may not be a good idea.

There's a good chance that you will want to sort the view in various ways. Even if there's one sort order you prefer, you don't want the DBMS to sort the result unnecessarily before you sort them again.

What you should do, however, is make sure that your view includes whatever columns you'll be sorting on later.

Table-Valued Functions

Views are a powerful tool, but there's one shortcoming: you can't change any of the values used in a view. For example, the `aupricelist` view has a hard-coded tax rate of 10%. A more flexible type of view would allow you to input your own tax rate. Such a view would then be called a **parameterized view**.

Parameterized views are not generally supported in SQL. Some DBMSs support functions which generate a virtual table, known as a **Table-Valued Function**, or TVF if you're in a hurry. This will give more or less the same result.

Of our popular DBMSs, only PostgreSQL and Microsoft SQL Server support a straightforward method of creating a TVF. We'll explore these two in the following discussion.

Most DBMSs allow you to create custom functions. The notable exception is SQLite, which does, however, allow you to create functions externally and hook them in.

A function which generates a single value at a time is called a **scalar** function. Built-in functions such as `lower()` and `length()` are scalar functions.

When creating a function, there is, in a sense, a contract. The function definition includes what input data is expected and what sort of data will be returned. If the input data doesn't fit, then don't expect a result.

A TVF works the same way: you define what input is expected, and you promise to return a table of results. Here, we'll create a more generic price list which allows you to tell it what the tax rate is, rather than hard-coding it.

To use the TVF, you use it like any virtual table:

```
SELECT *
FROM pricelist(15);
```

Here, the TVF is called `pricelist()` and the input parameter is 15, meaning 15%. The code should handle converting that to 0.15:

id	Title	pub	author	price	tax	inc
2078	The Duel	1811	Heinrich von Kleis ...	12.50	1.88	14.38
503	Uncle Silas	1864	J. Sheridan Le Fan ...	17.00	2.55	19.55
2007	North and South	1854	Elizabeth Gaskell	17.50	2.63	20.13
702	Jane Eyre	1847	Charlotte Brontë	17.50	2.63	20.13
1530	Robin Hood, The Pr ...	1862	Alexandre Dumas	12.50	1.88	14.38
1759	La Curée	1872	Émile Zola	16.00	2.40	18.40

~ 1070 rows ~

We'll do that for PostgreSQL and MSSQL next.

TVFs in PostgreSQL

The outline of a TVF in PostgreSQL looks like this:

```
CREATE FUNCTION pricelist(...)
RETURNS TABLE (...)
LANGUAGE plpgsql AS
$$
BEGIN
    RETURN QUERY
    SELECT ...
END
$$;
```

In this outline

- The function name `pricelist` includes the input parameter names and types.

- The function will return a TABLE structure with column names and types.

- The coding language is `plpgsql` which is PostgreSQL's standard coding language.

- The actual code is contained in one big string. Because there might be other strings in the code, the $$ at either end acts as an alternative delimiter.

- The code is then placed between `BEGIN` and `END`; in this case, it will return the results of a `SELECT` query.

Filling in the details, we can write

```
DROP FUNCTION IF EXISTS pricelist(taxrate decimal(4,2));
CREATE FUNCTION pricelist(taxrate decimal(4,2))
RETURNS TABLE (
    id int, title varchar, published int, author text,
    price decimal(5,2), tax decimal(4,2), inc decimal(5,2)
)
LANGUAGE plpgsql AS $$
BEGIN
    RETURN QUERY
    SELECT
        b.id, b.title, b.published,
        coalesce(a.givenname||' ','') || coalesce(othernames||' ','')
            || a.familyname AS author,
        b.price, b.price*taxrate/100 AS tax,
        b.price*(1+taxrate/100) AS inc
    FROM books as b LEFT JOIN authors a ON b.authorid=a.id
    WHERE b.price IS NOT NULL;
END; $$;
```

The output table is the most tedious part. In it, we have to list all of the column names and types we're expecting to generate.

As for the calculation, we've taken a user-friendly approach and allowed the tax rate to resemble the percentage we might have used in real life. We can't use %, especially as that has another meaning, but other than that, we can use the value. However, we then need to divide by 100 to get its real value.

TVFs in Microsoft SQL

Creating a TVF in Microsoft SQL is much simpler than in PostgreSQL. The outline of a TVF in Microsoft SQL looks like this:

```
GO
    CREATE FUNCTION pricelist(...) RETURNS TABLE AS
    RETURN SELECT ...
GO
```

There are two types of TVF in MSSQL. There is a more complex type, but the simpler type earlier is very similar to creating a view.

In this outline

- The function name `pricelist` includes the input parameter names and types.

- The function will return a `TABLE` structure.

- In the simple TVF, there is only a single `SELECT` statement, which is immediately returned as the result.

- The actual code is almost the same as for the view, except that it will include the value from the input parameter.

Filling in the details, we can write

```
DROP FUNCTION IF EXISTS pricelist;
GO
    CREATE FUNCTION pricelist(@taxrate decimal(4,2))
    RETURNS TABLE AS
        RETURN SELECT
            b.id, b.title, b.published,
            coalesce(a.givenname+' ','')
                + coalesce(othernames+' ','')
                + a.familyname AS author,
            b.price, b.price*@taxrate/100 AS tax,
            b.price*(1+@taxrate/100) AS inc
        FROM books as b JOIN authors a ON b.authorid=a.id
        WHERE b.price IS NOT NULL;
GO
```

The input parameter is called @taxrate. Actually, it's really called `taxrate`, but MSSQL uses the @ character to prefix all variables.

As with the PostgreSQL version, we've taken a user-friendly approach and allowed the tax rate to resemble the percentage we might have used in real life. We can't use %, especially as that has another meaning, but other than that, we can use the value. However, we then need to divide by 100 to get its real value.

What Can You Do with a View?

Given that a view is a virtual table, what sorts of things can be done with a view?

Convenience

The most immediate use of a view is as a convenient way of packaging a useful SELECT query. For example:

```
SELECT * FROM customerdetails;
SELECT * FROM aupricelist;
```

Both of the preceding views include joins, and one includes a number of calculations. It's much more convenient to use the saved view when you need it.

As an Interface

A second use of views is to present a consistent interface for existing data.

For example, when we refactored the `customers` table by referencing another table and dropping a few columns, we ran the risk of invalidating any other queries which depended on the old structure. By creating the `customerdetails` view, you have a new virtual table which can be read the same way as the old table.

It can also be handy if you're in the process of renaming or rearranging tables and columns. Suppose, for example, you're in the process of developing a new version of the `customers` table, with some of the following columns:

customerid	firstname	lastname	height_in_inches	au_phone	...
...	...	...	...	...	...

You can prepare for it with something like the following:

```
/* Notes

    ============================================================
    MSSQL:              Use + for concatenation
    Oracle, SQLite: Use substr(phone,2) instead of right()
    ============================================================ */
    --  CREATE VIEW newcustomers AS
    SELECT
        id AS customerid,
        givenname AS firstname, familyname AS lastname,
        cast(height/2.54 as decimal(3,1))
            AS height_in_inches,
        '+61' || right(phone,9) AS au_phone
        --  etc
    FROM customers;
```

This will give you something like

customerid	firstname	lastname	height_in_inches	au_phone
42	May	Knott	66.3	+61255509371
459	Rick	Shaw	67.3	+61370101040
597	Ike	Andy	60.2	[NULL]
186	Pat	Downe	69.3	+61870105900
352	Basil	Isk	61.6	+61255502503
576	Pearl	Divers	69.4	+61370107821

~ 303 rows ~

(The CREATE VIEW clause is commented out, because we're not really going to go ahead with this.)

This approach will also be useful if you're preparing data for an external application.

Working with External Applications

Although we've spent our time with a front-end client, that is not necessarily the destiny of the data. Often, we will use the data in an external application such as reporting software.

It makes the most sense to preprocess the data as much as possible before using it with an external application. In particular, joining tables or running subqueries is something which should be at the database end, not later.

Note that there are some things the database cannot do so readily. For example, there may be data formatting requirements or certain functions which are not available in the database server. For that, you will need to extract the raw data and process it in the external application.

Examples of working with external applications include

- Mail Merge with a word processor

- Working with Pivot Tables in a spreadsheet

- Working with reporting software

Such software typically has very limited ability in manipulating data, so it makes sense to do as much preprocessing as possible. When seen from the external software, your views will be perceived as single tables (though often they'll still indicate that they are actually views).

Caching Data and Temporary Tables

Some DBMSs offer a materialized view. This is a type of view which caches the data. The result is that repeated reading from the view doesn't (necessarily) require reprocessing.

When you're using a view, you may be putting an extra burden on the DBMS. If the view has some complex processing and joining to do, then every time you look at the view, you find yourself redoing the extra work.

Some DBMSs cheat a little by keeping a copy of the results for next time, as long as next time isn't too much later and nothing's changed in the associated tables. This copy is called a **cache**, and it's up to the DBMS whether to do this or not.

Some DBMSs support a materialized view, which formalizes the process. A materialized view has storage allocated to maintaining a copy of the results with the database. This is cheaper on processing, because the DBMS doesn't need to process the same data so often, but more expensive on storage. As usual, you may find that extra storage is cheaper than processing power.

Materialized views aren't widely supported and are sometimes limited in usefulness. However, you go a long way with temporary tables.

In principle, all SQL tables are temporary, in that it's always possible to drop a table—in SQL, as in life, nothing's truly permanent. However, a temporary table is one destined to be short-lived and will self-destruct when you close the session.

You can create a temporary table as you might a real table, but using the TEMPORARY prefix:

```
-- PostgreSQL, MariaDB/MySQL, SQLite
   CREATE TEMPORARY TABLE somebooks (
       id INT PRIMARY KEY,
       title VARCHAR(255),
       author VARCHAR(255),
       price DECIMAL(4,2)
   );

-- Oracle
   CREATE GLOBAL TEMPORARY TABLE somebooks (
       id INT PRIMARY KEY,
       title VARCHAR(255),
       author VARCHAR(255),
       price DECIMAL(4,2)
   );

-- MSSQL
   CREATE #somebooks (
       id INT PRIMARY KEY,
       title VARCHAR(255),
       author VARCHAR(255),
       price DECIMAL(4,2)
   );
```

Note

- Oracle distinguishes between GLOBAL and PRIVATE temporary tables and requires one of the two keywords. Private temporary tables also need to have special names.

- PostgreSQL allows you to use the GLOBAL and LOCAL keywords for the same purpose, but then ignores them; they recommend leaving them out.

- MSSQL uses hashes for global and private temporary tables: one hash (#) for private and two hashes (##) for global.

By "global," we mean that other uses of the database can access the temporary table. Private ones are, well, private to the session.

If you're in a desperate hurry, PostgreSQL and SQLite allow you to save time by writing TEMP instead of TEMPORARY. It probably took you more time to read this paragraph.

The temporary table in this example has a simple integer primary key. If you intend adding more data as you go, you might also use an autoincremented primary key.

Once you have created your temporary table, you can copy data into it using the SELECT statement. For example:

```
INSERT INTO somebooks(id,title,author,price)
SELECT id,title,author,price
FROM aupricelist
WHERE price IS NOT NULL;
```

The INSERT ... SELECT ... statement copies data into an existing table, temporary or permanent.

You can create a new table and populate it in one statement with the following statement:

```
-- PostgreSQL, MariaDB/MySQL, Oracle
   CREATE TABLE otherbooks AS
       SELECT id,title,author,price
       FROM aupricelist
       WHERE price IS NULL
   ;
```

```
--   PostgreSQL, SQLite
     SELECT id,title,author,price
     INTO TEMPORARY otherbooks
     FROM aupricelist
     WHERE price IS NULL;

--   MSSQL
     SELECT id,title,author,price
     INTO #otherbooks
     FROM aupricelist
     WHERE price IS NULL;
```

As you see, this statement takes one of two forms; PostgreSQL supports both.

Note that either form requires that you have permissions to create either a temporary or permanent table.

Remember, however, that the data is a copy, so it will go stale unless you update it.

Why would you want a temporary table? There's nothing in our sample database which could be regarded as in any way heavy-duty. However, in the real world, you might be working with a query which involves a huge number of rows, complex joins, filters and calculations, and sorting. This could end up taking a great deal of time and effort, especially if you're constantly regenerating the data.

The reasons you would use a temporary table rather than a view include

- It's more efficient to save previously generated results than it is to regenerate them. This is called **caching** the results.

- Sometimes, you *want* the data to be out of date, such as when you need to work with a **snapshot** of the data from earlier in the day.

  If you need to work with the snapshot at some point in the future, a temporary table may be too fleeting. Everything we've done will also apply to specially created permanent tables.

A database should never keep multiple copies of data. However, there are times when you need a temporary table for further processing, experimenting, or in transit to migrating data.

Computed Columns

Modern SQL allows you to add a column to a table which in principle shouldn't be in a table. A **computed column**, or calculated column, is an additional column which is based on some calculated value. When you think about it, that's the sort of thing you would do in a view.

Think of the computed column as embedding a mini-view in the table. It's particularly handy if you commonly use one calculation but don't want the overhead of a view. It can also be handy if you have the option to cache the results.

A computed column is a read-only virtual column. You can't write anything into the column, and, if it saves any data at all, it's a cached value to save the effort of recalculating it later. For example, you might store the full name of the customer as a convenience.

You can create a computed column when you create the table, or you can add it to the table after the event.

For example, suppose we want to add a shortened form of the `ordered` datetime column, with just the date. This will be handy for summarizing by day.

You can add the new column as follows:

```
-- PostgreSQL >= 12
    ALTER TABLE sales
    ADD COLUMN ordered_date date
        GENERATED ALWAYS AS (cast(ordered as date)) STORED;

-- MSSQL
    ALTER TABLE sales
    ADD ordered_date AS (cast(ordered as date)) PERSISTED;

-- MariaDB / MySQL
    ALTER TABLE sales
    ADD ordered_date date
        GENERATED ALWAYS AS (cast(ordered as date)) STORED;

-- SQLite>=3.31.0
    ALTER TABLE sales
    ADD ordered_date date
        GENERATED ALWAYS AS (cast(ordered as date)) VIRTUAL;
```

```
--  Oracle (STORED)
    ALTER TABLE sales
    ADD ordered_date date
        GENERATED ALWAYS AS (trunc(ordered));
```

As you see, most DBMSs use the standard GENERATE ALWAYS syntax. MSSQL, however, uses its own simpler syntax which doesn't specify the data type but infers it from the calculation.

You'll also notice different types of computed column:

- VIRTUAL columns are not stored and are recalculated. This is the default in MSSQL.

- STORED columns save a copy of the result and will only recalculate if the underlying value has changed.

  MSSQL calls this PERSISTED. In Oracle, it's the default. SQLite does support this as well, but only if you create the table that way; if you add the column later, it can only be VIRTUAL.

You can now fetch the data complete with virtual column:

```
SELECT * FROM sales;
```

This gives you

id	...	Ordered	shipped	ordered_date
39	...	2022-05-15 21:12:07.988741	2022-05-23	2022-05-15
40	...	2022-05-16 03:03:16.065969	2022-05-24	2022-05-16
42	...	2022-05-16 10:09:13.674823	2022-05-22	2022-05-16
43	...	2022-05-16 15:02:43.285565	[NULL]	2022-05-16
45	...	2022-05-16 16:48:14.674202	2022-05-28	2022-05-16
518	...	[NULL]	[NULL]	[NULL]
~ 5549 rows ~				

If you have the option, the better option is STORED or equivalent. It takes a little more space, but saves on processing later.

Summary

Much of your work will involve not only real tables but generated virtual tables. Virtual tables include

- A join

- A Common Table Expression or a table subquery

- A view or, in some cases, a Table Valued Function

- A temporary table.

Views

A view is a saved SELECT statement. It can be made as complex as you like and then fetched as a virtual table.

The benefits of views include

- They can be a convenient way of working with data.

- They can act as an interface to your data, particularly where the original or modified form doesn't match your requirements.

- They offer a simple table view of complex data when accessed from external applications.

Table Valued Functions

A Table Valued Function is a function which results in a virtual table. Support for TVFs is sketchy. Where it is supported, it allows you to feed parameters into the query, making it more flexible than a simple view.

Temporary Tables

There are times when it is better to store results rather than regenerate them every time. You can save them into a caching table.

The benefits include

- It's more efficient not to have to recalculate what will be the same results.

- You might want to work with a dated snapshot of your data.

If your cache is intended to be particularly short-lived, you might use a temporary table. A temporary table is one which will self-destruct at the end of the session.

Whether the caching table is temporary or permanent, you can copy data into it using a SELECT statement. You can also create a new table and copy into it in a single statement.

Computed Columns

In modern DBMSs, you can create virtual columns in a table which give the results of a calculation.

A VIRTUAL computed column will regenerate the value every time you fetch from the table. A STORED computed column, a.k.a. PERSISTED in MSSQL, will cache the results until other data has changed.

A computed column can be used for convenience. If it's a STORED column, it also has the benefit of saving on processing.

Coming Up

A SELECT statement doesn't have to be the end of the story. In some cases, it can be one step in a more complex story.

A **subquery** allows you to embed a SELECT statement inside a query. This can be used to fetch values from other tables or to use one table to filter another. It's particularly handy if you want to incorporate aggregate data in another query.

The next chapter will look at subqueries in more detail.

CHAPTER 7

Working with Subqueries and Common Table Expressions

Running a SELECT statement, assuming that there's no error, gives you a result. That result is a virtual table, and it will have rows and columns.

For our purposes, we are interested in three possible virtual tables:

- One row and one column: You get just one value, though technically it's still in a table. We'll call this a single value.

- One column and multiple rows: When the time comes, we'll call this a list.

- Multiple rows and multiple columns: In this context, a single row with multiple columns counts as the same sort of thing. This is more like the sort of thing we think about when talking about virtual tables.

Of course, that result may be empty, but that's treated as NULLs.

You'll get these types of results from the following examples.

For example, one row and one column:

```
SELECT id FROM books WHERE title='Frankenstein';
```

which is a single value:

Id
392

© Mark Simon 2023
M. Simon, *Leveling Up with SQL*, https://doi.org/10.1007/978-1-4842-9685-1_7

One column and multiple rows:

```
SELECT email FROM customerdetails WHERE state='VIC';
```

which gives a list:

Email
xavier.money17@example.net
anne.onymous262@example.net
bess.twishes26@example.net
judy.free93@example.net
peter.off415@example.com
moe.grass360@example.com
~ 64 rows ~

Multiple rows and multiple columns:

```
SELECT givenname, familyname, email
FROM customerdetails WHERE state='VIC';
```

which gives a virtual table:

givenname	Familyname	email
Xavier	Money	xavier.money17@example.net
Anne	Onymous	anne.onymous262@example.net
Bess	Twishes	bess.twishes26@example.net
Judy	Free	judy.free93@example.net
Peter	Off	peter.off415@example.com
Moe	Grass	moe.grass360@example.com
~ 64 rows ~		

That last category, the virtual table, could also be the result of a very broad query such as SELECT * FROM customerdetails. It all works the same way.

Any of these results, depending on the context, can be used in a subsequent query where a single value, a list, or a (virtual) table might have been expected. For example, using a single value:

```
SELECT *
FROM saleitems
WHERE bookid=(SELECT id FROM books WHERE title='Frankenstein');
```

Here, the single value query is wrapped inside parentheses and used the way you would if you already knew the value of the bookid you're matching:

id	Saleid	bookid	quantity	price
7234	2873	392	3	18.50
14875	5907	392	2	18.50
11183	4448	392	2	18.50
1312	517	392	2	18.50
9956	3948	392	2	18.50
12636	5012	392	2	18.50
~ 14 rows ~				

Or you could be expecting some sort of list:

```
SELECT *
FROM books
WHERE authorid IN (
    SELECT id FROM authors WHERE born BETWEEN '1700-01-01' AND '1799-12-31'
);
```

The IN operator expects a list of values, which we get from the one column in the nested SELECT statement:

id	authorid	Title	published	price
2078	765	The Duel	1811	12.50
2243	715	Vrijmoedige Verhalen; ee …	1831	[NULL]
532	628	Elective Affinities	1809	11.50
1608	420	The Autobiography of Ben …	1791	18.50
2303	633	The Metaphysics of Moral …	1797	12.00
1963	529	A Treatise of Human Natu …	1740	18.50

~ 256 rows ~

The nested SELECT statement is called a **subquery**. A subquery is a SELECT statement which is used as part of another query. Conceptually, you use the SELECT query to fetch one or more results, which in turn will be used in another query.

Subqueries can be used in the SELECT, WHERE, FROM, and even ORDER BY clauses. When they are used, the result must be compatible with the context. For example:

- A subquery in a SELECT clause must always return a single value, as would be expected from any other calculation. This can be called a **scalar** (single-value) subquery.

  Some modern DBMSs are now adding support for returning values for more than one column, but it's not widely supported for now.

- A subquery in the WHERE clause must return a single value when used with a comparison operator, or a single column when used with an IN() expression.

- A subquery in the FROM clause must return a table of results.

  This type of subquery is sometimes called an inline view.

Some uses of subqueries include

- Looking up related data from another table (SELECT)

- Looking up a value or set of values to act as a filter (WHERE)

- Generating a derived virtual table (FROM)

The whole thing with subqueries is that with a subquery you can combine multiple parts to make a more complex query:

- Using a subquery, you'll be able to get data from other tables.

- You can use subqueries to preprocess data to be used in the main query.

- Subqueries will allow you to combine aggregate and non-aggregate queries, which you can't normally do in single queries.

Subqueries do have a cost, however, as you will be running two or more queries for the price of one—sometimes, even hundreds of queries for the price of one. We'll have a look at that later when we compare subqueries to alternative techniques.

You will no doubt have seen some subqueries before now, certainly in some of the earlier chapters in this book. In this chapter, we'll have a closer look at how subqueries work and what we can do with them.

Correlated and Non-correlated Subqueries

Subqueries can be **correlated** or **non-correlated**. A correlated subquery includes a reference to the main query. A non-correlated subquery is evaluated independently from the main query.

Here's an example of a non-correlated subquery:

```
-- Books by Female Authors
SELECT *
FROM books
WHERE authorid IN(
    SELECT id FROM authors WHERE gender='f'
);
```

You'll get a result like this:

id	authorid	Title	published	price
2007	99	North and South	1854	17.50
702	547	Jane Eyre	1847	17.50

(continued)

id	authorid	Title	published	price
95	701	Silas Marner	1861	18.50
983	211	East Lynne	1861	16.00
678	547	Tales of Angria	1839	14.50
1255	608	Discourse on the Origin ...	1755	19.00

~ 165 rows ~

Here's another which uses an aggregate in the subquery:

```
-- Oldest Customers
   SELECT *
   FROM customers
   WHERE dob=(SELECT min(dob) FROM customers);
```

This gives you

id	givenname	familyname	...	dob	...
92	Nan	Keen	...	1943-05-18	...
392	Daisy	Chain	...	1943-05-18	...

(You'll note that there's more than one oldest customer, because they happen to be born on the same day. It happens.)

In both cases, the subquery is evaluated once, and the results are used in the main query. The result may be a list, as in the female authors, or a single value as in the oldest customer.

A non-correlated subquery is independent of the main query. If you highlight the subquery alone and run it, you'll get a result.

Here's an example of a correlated subquery:

```
-- Book Authors (yes, there's another way to do this)
   SELECT
       id, title, (
           SELECT coalesce(givenname||' ','')
```

```
        || coalesce(othernames||' ',''))
        || familyname
        FROM authors
        WHERE authors.id=books.authorid
    ) AS author
FROM books;
--  MSSQL
SELECT
    id, title, (
        SELECT coalesce(givenname+' ',''
        + coalesce(othernames+' ','')
        + familyname
        FROM authors
        WHERE authors.id=books.authorid
    ) AS author
FROM books;

--  Oracle
SELECT
    id, title, (
        SELECT ltrim(givenname||' ')
        ||ltrim(othernames||' ')
        ||familyname
        FROM authors
        WHERE authors.id=books.authorid
    ) AS author
FROM books;
```

This gives you the following:

id	title	author
2078	The Duel	Heinrich von Kleist
503	Uncle Silas	J. Sheridan Le Fanu
2007	North and South	Elizabeth Gaskell

(continued)

id	title	author
702	Jane Eyre	Charlotte Brontë
1530	Robin Hood, The Prince of Thieves	Alexandre Dumas
1759	La Curée	Émile Zola
~ 1201 rows ~		

In this case, the subquery is evaluated once for every row. Look at the subquery in the first example earlier, spread out to be more readable:

```
(
    SELECT
        coalesce(givenname||' ','')
        || coalesce(othernames||' ','')
        || familyname
    FROM authors
    WHERE authors.id=books.authorid
)
```

The SELECT clause is expecting a single value for the author column, and so the subquery should deliver a single value, which it does. You can't use multiple columns in this context, so you need to concatenate the names to give the single value.

Just as importantly, you can't have multiple rows either. Here, the WHERE clause filters the result to a single row, where the id matches the authorid in the main query: WHERE authors.id=books.authorid.

For every row in the books table, the subquery runs again to match the next authorid.

If there's no match, the subquery comes back with a NULL.

You can recognize a correlated subquery by the fact that the query references something from the main query. As a result, you can't highlight the subquery and run it alone, because it needs that reference to be complete.

Incidentally, note the WHERE clause in the subquery. In a sense, it's overqualified, and we could have used this: WHERE id=authorid. This is in spite of the fact that an id column appears in both the subquery and the main query.

When the subquery is evaluated, column names will be defined from the inside out. For the id column, there's one in the inner authors table, so SQL doesn't bother to notice that there's also one in the outer books table. For the authorid column, there isn't one in the authors table, so it falls through the one in the books table.

That's how it works in SQL, but it's probably better to qualify the columns as we did in this example to minimize confusion for us humans.

As a rule, a correlated subquery is an expensive operation because it's reevaluated so often. That doesn't mean you shouldn't use one, just that you should consider the alternatives, if there are any. You don't generally get to choose which type of subquery you will need, but it will help in deciding whether there's a better alternative.

Subqueries in the SELECT Clause

As we saw, a subquery in a SELECT clause is expected to return a single value for each row. This can be tedious if, say, you want more than one value from the same external table with a correlated subquery:

```
SELECT
        id, title, (
            SELECT coalesce(givenname||' ','')
            || coalesce(othernames||' ','')
            || familyname
            FROM authors
            WHERE authors.id=books.authorid
        ) AS author,
        (SELECT born FROM authors
            WHERE authors.id=books.authorid) AS born,
        (SELECT born FROM authors
            WHERE authors.id=books.authorid) AS died
    FROM books;
```

Apart from being tedious, it's also expensive, and, of course, there's a better way to do it, using a join:

```
SELECT
    id, title,
    coalesce(givenname||' ','')
```

```
    || coalesce(othernames||' ','')
    || familyname AS author,
    born, died
FROM books AS b LEFT JOIN authors AS a ON b.authorid=a.id;
--  Oracle
--  FROM books b LEFT JOIN authors a ON b.authorid=a.id;
```

In fact, you'll probably find that a correlated subquery is often best replaced by a join. There's also some cost in the join, but after that, the rest of the data is free.

On the other hand, if the subquery is non-correlated, then it's not so expensive. For example, here's the difference between customers' heights and the average height:

```
SELECT
    id, givenname, familyname,
    height,
    height-(SELECT avg(height) FROM customers) AS diff
FROM customers;
```

You should see something like this:

id	givenname	familyname	height	Diff
42	May	Knott	168.5	-2.34
459	Rick	Shaw	170.9	0.06
597	Ike	Andy	153.0	-17.84
186	Pat	Downe	176.0	5.16
352	Basil	Isk	156.4	-14.44
576	Pearl	Divers	176.3	5.46

~ 303 rows ~

Even though the average is involved in a calculation in every row, it's only calculated once in the non-correlated subquery.

By the way, there's an alternative way to do the preceding query involving **window functions**, which we'll look at in Chapter 8. However, in this case, there's not much difference in the result.

You'll have noticed that, in this case, the subquery references the same table as the main query. That doesn't make it a correlated subquery, as it doesn't reference the actual *rows* in the main query. You can verify that if you highlight the subquery and run it by itself—it will work.

The subquery in this example was an aggregate query. You can also use an aggregate in a correlated query. Here's a way of generating a running total:

```
-- Oracle: FROM sales ss
SELECT
    id, ordered, total,
    (SELECT sum(total) FROM sales AS ss
    WHERE ss.ordered<=sales.ordered) AS running_total
FROM sales
ORDER BY id;
```

You'll get something like this:

id	ordered	total	running_total
1	2022-05-04 21:53:55.165107	43.00	43.00
2	2022-05-05 12:39:41.438631	54.50	97.50
3	2022-05-05 17:48:08.433387	96.00	193.50
4	2022-05-07 08:29:35.61573	17.50	321.50
5	2022-05-07 13:10:25.441528	63.00	384.50
6	2022-05-06 17:23:38.261261	18.00	211.50

~ 5549 rows

We've had to alias the table in the subquery to something like ss (subsales?) to distinguish it from the same table in the main query. That's so that the expression ss.ordered<=sales.ordered can reference the correct tables.

Here, the subquery calculates the sum of the totals up to and including the current sale, ordered by the ordered column.

You possibly noticed that the query took a little while to run. As we noted, a correlated subquery is costly, and one which involves aggregates is especially costly. Fortunately, there's also a window function for that, as we'll see in the next chapter.

Subqueries in the WHERE Clause

A subquery can also be used to filter your data.

Again, you may find an alternative to subqueries, such as JOINs, but one compelling use case is when the subquery is an aggregate query.

Here are some cases where the subquery makes the point clearly and simply.

Subqueries with Simple Aggregates

One place where a subquery is useful is when you need to mix aggregate and non-aggregate queries. For example, if you want to find the oldest customer, you'll need to do it in two steps:

1. Use an aggregate query to find the earliest date of birth.

2. Use the aggregate result as a filter.

Here, you see the aggregate query used as a subquery:

```
SELECT *
FROM customers
WHERE dob=(SELECT min(dob) FROM customers);
```

You can also do the same to find customers shorter than the average:

```
SELECT *
FROM customers
WHERE height<(SELECT avg(height) FROM customers);
```

In both cases, the aggregate query was on the same table as the main query. You might have thought that you could use an expression like WHERE dob=min(dob) or WHERE height<avg(height), but it wouldn't work; aggregates are calculated *after* the WHERE clause.

Big Spenders

Suppose you want to identify your "big spenders"—the customers who have spent the highest amounts. For that, you will need data from the customers and sales tables.

Here, we'll use subqueries as part of a multistep process.

To begin with, you'll want to identify what you regard as large purchases:

```
SELECT * FROM sales WHERE total>160;
```

You'll get something like this:

id	customerid	total	...	ordered_date
80	32	168.00	...	2022-05-22
216	13	160.50	...	2022-06-11
483	59	176.50	...	2022-07-11
726	68	173.00	...	2022-08-02
823	86	165.50	...	2022-08-09
891	140	162.50	...	2022-08-16

~ 35 rows ~

In here, we're only interested in the `customerid`, which we'll use to select from the customers table:

```
SELECT *
FROM customers
WHERE id IN(SELECT customerid FROM sales WHERE total>160);
```

This gives you

id	...	familyname	givenname	...
42	...	Knott	May	...
58	...	Ting	Jess	...
91	...	North	June	...
140	...	Byrd	Dicky	...
40	...	Face	Cliff	...
141	...	Rice	Jasmin	...

~ 32 rows ~

Note that the IN operator requires a *list* of values. In a subquery, this is a single column of values.

Note also that you may have few results than in the previous query; that would be if some of the customer ids appear more than once.

SQL also has an ANY operator which will do the same job:

```
SELECT *
FROM customers
WHERE id=ANY(SELECT customerid FROM sales WHERE total>=160);
```

You could also have used a JOIN:

```
SELECT DISTINCT customers.*
FROM customers JOIN sales ON customers.id=sales.customerid
WHERE sales.total>=160;
```

To recreate what we had in the previous query, we've qualified the star (customers.*) and used DISTINCT to remove duplicates of customers who may have appeared in the list more than once.

The advantage of using a join is that you can also get sales data for the asking, so this gives a slightly richer result:

```
SELECT *
FROM customers JOIN sales ON customers.id=sales.customerid
WHERE sales.total>=160;
```

Here, we've removed the DISTINCT and the customers., so you'll get a lot of data:

id	...	familyname	givenname	...	total	...	ordered_date
32	...	Cue	Barbie	...	168.00	...	2022-05-22
13	...	Fine	Marty	...	160.50	...	2022-06-11
59	...	Don	Leigh	...	176.50	...	2022-07-11
68	...	Stein	Phyllis	...	173.00	...	2022-08-02
86	...	Fied	Molly	...	165.50	...	2022-08-09
140	...	Byrd	Dicky	...	162.50	...	2022-08-16

~ 35 rows ~

To find customers with large *total* sales will require an aggregate subquery:

```
SELECT *
FROM customers
WHERE id IN(
```

```
SELECT customerid FROM sales
GROUP BY customerid HAVING sum(total)>=2000
);
```

This will give you

id	...	familyname	givenname	...
42	...	Knott	May	...
58	...	Ting	Jess	...
26	...	Twishes	Bess	...
91	...	North	June	...
69	...	Mentary	Rudi	...
140	...	Byrd	Dicky	...

~ 57 rows ~

Last Orders, Please

Here again, we'll use an aggregate subquery, this time to fetch the last order of each customer.

First, we'll need to fetch the last date and time for each customer:

```
SELECT max(ordered) FROM sales GROUP BY customerid
```

This will give us a list of one datetime for each customer:

Max
2023-05-15 00:46:00.864446
2023-05-25 00:42:26.783461
2023-05-16 05:27:53.810977
2023-05-06 01:40:02.346894
2023-05-19 07:41:25.104524
2023-05-07 19:01:06.756387
~ 269 rows ~

We'll use this list to fetch the matching orders:

```
SELECT * FROM sales
WHERE ordered IN(SELECT max(ordered) FROM sales GROUP BY customerid);
```

This will give us a list of sales:

id	customerid	total	...	ordered_date
6168	42	121.22	...	2023-05-28
4209	287	50.50	...	2023-03-03
4542	26	11.00	...	2023-03-18
4793	368	56.00	...	2023-03-28
4939	282	39.00	...	2023-04-03
4953	395	75.50	...	2023-04-03

~ 266 rows ~

If you count the rows, you may find that the main query returned fewer rows than the subquery. That would happen if there were some NULL ordered datetimes. At some point, we should learn to ignore these, either by filtering them out or removing them altogether.

The question is, why weren't those sales included in the full query? And the answer is that it's all about the IN() operator.

Remember in Chapter 3, we discussed the NOT IN quirk. The discussion also applies to a plain IN. The NULL datetimes in the subquery would result in the equivalent of testing WHERE ordered=NULL, which, as we all know, always fails.

Now that we have sales for each customer, it's a simple matter to join that to the customers table to get more details:

```
SELECT *
FROM sales JOIN customers ON sales.customerid=customers.id
WHERE ordered IN(SELECT max(ordered) FROM sales GROUP BY customerid);
```

You'll get something like

id	cid	total	...	od	id	email	...
6168	42	121.22	...	2023-05-28	42	may.knott61@example.net	...
4209	287	50.50	...	2023-03-03	287	judy.free287@example.com	...
4542	26	11.00	...	2023-03-18	26	bess.twishes26@example.net	...
4793	368	56.00	...	2023-03-28	368	sharon.sharalike368@example.net	...
4939	282	39.00	...	2023-04-03	282	howard.youknow282@example.com	...
4953	395	75.50	...	2023-04-03	395	holly.day395@example.net	...

~ 266 rows ~

You can now extract any customer or sales data you might want to work with.

Duplicated Customers

We've seen in Chapter 2 how to find duplicates. Suppose, for example, you want to find duplicate customer names:

```
SELECT
    givenname||' '||familyname AS fullname,
    --  MSSQL: givenname+' '+familyname AS fullname,
    count(*) as occurrences
FROM customers
GROUP BY familyname, givenname
HAVING count(*)>1;
```

You get

fullname	Occurrences
Judy Free	2
Annie Mate	2
Mary Christmas	2

(continued)

251

fullname	Occurrences
Ken Tuckey	2
Corey Ander	2
Ida Dunnit	2
Paul Bearer	2
Terry Bell	2

Remember, having the same name doesn't necessarily mean they're duplicates. It's probably just a coincidence.

We've concatenated the name because of what we're going to do in the next step.

The problem with aggregate queries is that you can only select what you're grouping, so we can't see the rest of the customer details. Any attempt to include them would destroy the aggregate.

We can, however, use the duplicate query as a subquery to filter the customers table:

```
/*  Note
    ================================================
    MSSQL: Use givenname+' '+familyname
    ================================================ */
SELECT *
FROM customers
WHERE givenname||' '||familyname IN (
    SELECT givenname||' '||familyname FROM customers
    GROUP BY familyname, givenname
    HAVING count(*)>1
);
```

This will give us the rest of the customer details. The reason we had to concatenate the customers' names is that you can only have a single column in the IN() expression.

Subqueries in the FROM Clause

A subquery can also generate a virtual table, which is useful when you need to prepare data before actually querying it.

For example, suppose you want to look at your books in price groups. You can create a simple query like this:

```
SELECT
    id, title,
    CASE
        WHEN price<13 THEN 'cheap'
        WHEN price<=17 THEN 'reasonable'
        WHEN price>17 THEN 'expensive'
    END AS price_group
FROM books;
```

You'll get something like this:

id	title	price_group
2078	The Duel	Cheap
503	Uncle Silas	Reasonable
2007	North and South	Expensive
702	Jane Eyre	Expensive
1530	Robin Hood, The Prince of Thieves	Cheap
1759	La Curée	Reasonable
~ 1201 rows ~		

Now, suppose you want to summarize the table. The problem is that you can't do this:

```
--  This doesn't work:
    SELECT
    --  id, title,
        CASE
            WHEN price<13 THEN 'cheap'
            WHEN price<=17 THEN 'reasonable'
            WHEN price>17 THEN 'expensive'
        END AS price_group,
```

253

```
       count(*) as num_books
  FROM books
  GROUP BY price_group;
```

We've commented out the columns we're not grouping, but it still won't work because of that pesky clause order thing: the alias `price_group` is created in the SELECT clause which comes *after* the GROUP BY clause, so it's not available for grouping. Of course, you can then reproduce the calculation in the GROUP BY clause:

```
-- This works, but ...
  SELECT
  --   id, title,
      CASE
          WHEN price<13 THEN 'cheap'
          WHEN price<=17 THEN 'reasonable'
          WHEN price>17 THEN 'expensive'
      END AS price_group,
      count(*) as num_books
  FROM books
  GROUP BY CASE
          WHEN price<13 THEN 'cheap'
          WHEN price<=17 THEN 'reasonable'
          WHEN price>17 THEN 'expensive'
      END;
```

but you really don't want to go there.

One solution is to put the original SELECT statement in a subquery:

```
SELECT price_group, count(*) AS num_books
FROM (
    SELECT
        id, title,
        CASE
            WHEN price<13 THEN 'cheap'
            WHEN price<=17 THEN 'reasonable'
            WHEN price>17 THEN 'expensive'
        END AS price_group
```

```
    FROM books
) AS sq      --  Oracle: ( ... ) sq
GROUP BY price_group;
```

This gives a meaningful result:

price_group	num_books
expensive	320
[NULL]	105
reasonable	467
cheap	309

Remember that the default fall through for the CASE expression is NULL. Those books which are unpriced will end up in the NULL price group. Depending on the DBMS, you'll see this somewhere in the result set as a separate group.

Remember that a SELECT statement generates a virtual table. As such, it can be used in a FROM clause in the form of a subquery.

Note that there's a special requirement for a FROM subquery: it must have an alias, even if you've no plans to use it. We have no special plans here, so it's just called sq ("SubQuery") for no particular reason. If you want to, say, join the subquery with another table or virtual table, then the alias will be useful.

Nested Subqueries

A subquery is a SELECT statement with its own FROM clause. In turn, that FROM clause might be from another subquery. If you have a subquery within a subquery, it's a **nested** subquery.

For example, let's look at duplicate customer names again. You can find candidates with the following aggregate query:

```
SELECT familyname, givenname
FROM customers
GROUP BY familyname, givenname HAVING count(*)>1;
```

They're just the names. Suppose you want more details. For that, you can join the customers table with the preceding query:

```
SELECT
    c.id, c.givenname, c.familyname, c.email
FROM customers AS c JOIN (
    SELECT familyname, givenname
    FROM customers
    GROUP BY familyname, givenname HAVING count(*)>1
) AS n ON c.givenname=n.givenname AND c.familyname=n.familyname;
```

We've seen something like this before. You'll now get the candidate customers:

id	givenname	familyname	Email
429	Corey	Ander	corey.ander429@example.net
287	Judy	Free	judy.free287@example.com
90	Ida	Dunnit	ida.dunnit90@example.net
488	Ken	Tuckey	ken.tuckey488@example.net
174	Paul	Bearer	paul.bearer174@example.com
505	Annie	Mate	annie.mate505@example.com

~ 16 rows ~

We've aliased the customers table to c for convenience (don't forget no AS in Oracle), and the subquery needs a name anyway, so we've called it n. In the SELECT clause, we've just fetched the id, the names, and the email address.

Now, let's combine this in another aggregate query, which will give us one row per name, and combine the other details:

```
SELECT
    givenname, familyname,
    -- PostgreSQL, MSSQL:
        string_agg(email,', ') AS email,
        string_agg(cast(id AS varchar(3)),', ') AS ids
    -- MariaDB/MySQL:
        group_concat(email SEPARATOR ', ') AS email,
```

```
    group_concat(cast(id AS varchar(3)) SEPARATOR ', ')
        AS ids
  -- SQLite:
    group_concat(email,', ') AS email,
    group_concat(cast(id AS varchar(3)),', ') AS ids
  -- Oracle:
    listagg(email,', ') AS email,
    listagg(cast(id AS varchar(3)),', ') AS ids
FROM (  --  previous SELECT as subquery
    SELECT c.id, c.givenname, c.familyname, c.email
    FROM customers AS c JOIN (
        SELECT familyname, givenname
        FROM customers
        GROUP BY familyname, givenname HAVING count(*)>1
    ) AS n ON c.givenname=n.givenname AND
        c.familyname=n.familyname
) AS sq
GROUP BY familyname, givenname;
```

Now you'll get something like

givenname	familyname	Email	ids
Corey	Ander	corey.ander429@e …, corey.ander85@ex …	429, 85
Paul	Bearer	paul.bearer174@e …, paul.bearer482@e …	174, 482
Terry	Bell	terry.bell402@ex …, terry.bell295@ex …	402, 295
Mary	Christmas	mary.christmas46 …, mary.christmas59 …	465, 594
Ida	Dunnit	ida.dunnit504@ex …, ida.dunnit90@exa …	504, 90
Judy	Free	judy.free93@exam …, judy.free287@exa …	93, 287

~ 8 rows ~

Note that we've included variations on the string_agg() functions as described in Chapter 5. Also, we've had to cast the id as a string so that it can be aggregated as one.

If you think this last example is getting a little hard to read, well, that's one of the problems with nested subqueries. Thankfully, there's an alternative in Common Table Expressions, which we'll look at soon.

Using WHERE EXISTS (Subquery)

You have already seen subqueries in the WHERE clause to filter your query. So far, the subquery has returned either a single value or a single column for use in an IN() expression.

You can also filter a set of data using WHERE EXISTS (...). It looks a little like this:

```
SELECT ...
FROM ...
WHERE EXISTS(subquery);
```

The subquery will either return a result or not. If it does, then the WHERE EXISTS is satisfied, and the row is passed; if it doesn't, then the WHERE EXISTS isn't satisfied, and the row will be filtered.

For example, you can test the idea with the following statement:

```
-- PostgreSQL, MSSQL, SQLite
   SELECT * FROM authors
   WHERE EXISTS (SELECT 1 WHERE 1=1);
-- MariaDB/MySQL, Oracle
   SELECT * FROM authors
   WHERE EXISTS (SELECT 1 FROM dual WHERE 1=1);
```

Since 1=1 is always true, you'll get all of the rows from the authors table.

Although you would normally only use FROM dual with Oracle, MariaDB and MySQL also support this. In this case, MariaDB and MySQL don't like the WHERE clause without a FROM, so we've thrown it in to keep them happy.

Similarly, you can return nothing:

```
-- PostgreSQL, MSSQL, SQLite
   SELECT * FROM authors
   WHERE EXISTS (SELECT 1 WHERE 1=0);
-- MariaDB/MySQL, Oracle
   SELECT * FROM authors
   WHERE EXISTS (SELECT 1 FROM dual WHERE 1=0);
```

The subquery is in a special position in that it doesn't matter what columns are actually being selected: what matters is that there is or isn't a row. That's why we've included a dummy SELECT 1.

You can also choose SELECT NULL or even SELECT 1/0. The former would give the (false) impression that we're looking for nothing, and the latter would have resulted in an error if run by itself. It's tempting to take it more seriously by selecting a more meaningful value, but there's no need.

WHERE EXISTS with Non-correlated Subqueries

Clearly, the previous examples will give you all or nothing, and that's the sort of thing you can expect from a non-correlated subquery, such as

```
SELECT * FROM authors
WHERE EXISTS (SELECT 1/0 FROM books WHERE price<15);
```

The subquery selects *some* rows, which is enough to satisfy the WHERE clause, so you'll get *all* the authors. If you had tried WHERE price<0, then you'd get *none* of the authors.

WHERE EXISTS with Correlated Subqueries

Using WHERE EXISTS is more interesting if you use a correlated subquery. Here, the test will be evaluated for every row, so some rows will pass and some won't.

For example, if you want to find all the authors whose books are in the books table, you can use

```
SELECT * FROM authors
WHERE EXISTS (
    SELECT 1 FROM books WHERE books.authorid=authors.id
);
```

You'll get something like

id	givenname	...	familyname	...	Home
464	Ambrose	...	Bierce	...	Meigs County, Ohio
858	Alexander	...	Ostrovsky	...	Moscow
525	Francis	...	Beaumont	...	Grace-Dieu, Leicestershire
488	Bashou	...	Matsuo	...	Matsuo Kinsaku

(continued)

id	givenname	...	familyname	...	Home
703	Friedrich	...	Engels	...	Barmen
722	Stanley	...	Waterloo	...	St. Clair, Mich.
~ 443 rows ~					

Here, the subquery looks for a row where `books.authorid` matches `authors.id`, and if there is such a row (as there is with most authors), the author row will be returned.

WHERE EXISTS vs. the IN() Expression

In the previous example, you can also get the same result with the `IN()` expression:

```
SELECT * FROM authors
WHERE id IN(SELECT authorid FROM books);
```

This variation is, of course, simpler. However, it's quite likely that, on the inside, SQL does exactly the same thing, so how you write it is really a matter of taste.

On the other hand, if you're looking for authors *without* books (in our catalogue), then it's a different matter.

This *won't* work:

```
SELECT * FROM authors
WHERE id NOT IN(SELECT authorid FROM books);
```

Well, technically, it *will* work, but not the way we would have wanted. Recall again from Chapter 3 the "NOT IN quirk." Since there are some NULLs in the `authorid` column, the `NOT IN` operator eventually evaluates something like `... AND id=NULL AND ....` The `id=NULL` always fails, and the `... AND ...` combines that failure with the rest and causes the whole expression to fail.

Using `WHERE NOT EXISTS` will, however, work:

```
SELECT * FROM authors
WHERE NOT EXISTS (
    SELECT 1 FROM books WHERE books.authorid=authors.id
);
```

That's because WHERE EXISTS evaluates rows, not values:

id	givenname	...	familyname	...	Home
479	C.E.	...	Koetsveld	...	Rotterdam
874	Henry	...	Savery	...	Somerset, England
429	Oliné	...	Keese	...	[NULL]
35	James	...	Lowell	...	Cambridge, Massachusetts
148	Demetrius	...	Boulger	...	[NULL]
922	Robert	...	Ingersoll	...	Dresden, New York
~ 45 rows ~					

You won't see WHERE EXISTS much in the wild, since you can generally do the same thing with either a join or the IN operator. However, there are times where it has an advantage or is more intuitive. That's especially because WHERE EXISTS can be more expressive and particular when NOT IN doesn't work.

LATERAL JOINS (a.k.a. CROSS APPLY) and Friends

This feature is not available in SQLite. Neither is it available in MariaDB, though it *is* available in MySQL.

Still, if you're working with one of these DBMSs, you might want to see what it's all about. In any case, there's an alternative for most things, especially with Common Table Expressions, in the next section.

SQLite does, however, have an interesting quirk with the WHERE clause, which you'll see if you hang around.

If you try the following query:

```
SELECT
    id, title,
    price, price*0.1 AS tax, price+tax AS inc
FROM books;
```

It won't work. That's because each column is independent of the rest. You can't use an alias as part of another calculation in the SELECT clause. We got around this by calculating the inc column separately: price*1.1 AS inc.

It gets worse if you try something like this:

```
SELECT
    id, title,
    price, price*0.1 AS tax
FROM books
WHERE tax>1.5;
```

Here, the problem is that the SELECT clause is evaluated *after* the WHERE clause, so the aliased calculation for tax isn't available yet in the WHERE clause. Again, we could recalculate the value in the WHERE clause: WHERE price*1.1>1.5.

Except with SQLite. You can indeed use aliases in the WHERE clause and also in the GROUP BY clause.

Finally, if, for example, you want to get multiple columns from a subquery in the SELECT clause, this won't work either:

```
SELECT
    id, title,
    (SELECT givenname, othernames, familynames
    FROM authors WHERE authors.id=books.authorid)
FROM books
WHERE tax>1.5;
```

A subquery in the SELECT clause can only return one value, which is all right if you concatenate the names and then return the result. Otherwise, you're stuck with three subqueries, which is both costly and tedious.

SQL can solve this by applying a subquery to each row. This is called a LATERAL JOIN in some DBMSs, or an APPLY in some others.

Adding Columns

In the first two examples earlier, you can use an expression like this:

```
-- PostgreSQL, MySQL (not MariaDB)
   SELECT
       id, title,
       price, tax, inc
   FROM
       books
       JOIN LATERAL(SELECT price*0.1 AS tax) AS sq ON true
       JOIN LATERAL(SELECT price+tax AS inc) AS sq2 ON true
   WHERE tax>1.5
   ;
-- MSSQL
   SELECT
       id, title,
       price, tax, inc
   FROM books
       CROSS APPLY (SELECT price*0.1 AS tax) AS sq
       CROSS APPLY (SELECT price+tax AS inc) AS sq2
   WHERE tax>1.5
   ;

-- Oracle (No AS in subquery name)
   SELECT
       id, title,
       price, tax, inc
   FROM books
       CROSS APPLY (SELECT price*0.1 AS tax FROM dual) sq
       CROSS APPLY (SELECT price+tax AS inc FROM dual) sq2
   WHERE tax>1.5
   ;
```

You'll get something like this:

id	Title	price	tax	inc
503	Uncle Silas	17.00	1.700	18.700
2007	North and South	17.50	1.750	19.250
702	Jane Eyre	17.50	1.750	19.250
1759	La Curée	16.00	1.600	17.600
205	Shadow: A Parable	17.50	1.750	19.250
1702	Philaster	17.50	1.750	19.250

~ 525 rows ~

Note

- The subquery must be given an alias, even though it's not used.

- PostgreSQL, MySQL, and MSSQL allow you to put the column aliases in the subquery aliases instead: (SELECT price*0.1) AS sq(tax). Not Oracle.

- The example for PostgreSQL and MySQL uses the dummy condition ON true. MySQL will allow you to leave this out, but PostgreSQL requires it.

Note in particular that the second subquery will happily calculate the expression price+tax AS inc. This is because the subqueries are evaluated one after the other, so the expressions can accumulate.

The LATERAL or CROSS APPLY subquery is applied to every row of the main query. In principle, that could be pretty expensive, but, as it turns out, it's not so bad. It's particularly useful if you need to include a series of intermediate steps in a more complex calculation—it's easy to understand and easy to maintain.

SQL also has a type of join called CROSS JOIN. In a cross join, each row of one table is joined with each row of the other table. This result is also known as a Cartesian product. That's a lot of combinations, and it's usually not what you want.

A CROSS APPLY is not the same thing, though it is a type of join. It's closer to an OUTER JOIN.

You'll see a use for a cross join later when we cross join with a single row virtual table.

Multiple Columns

As we noted, SQL won't let you fetch multiple columns from a single subquery in the SELECT clause, because everything in the SELECT clause is supposed to be scalar—a single value.

However, you can fetch multiple columns if the context is table-like, such as in the FROM clause. For example:

```
-- PostgreSQL, MySQL (Not MariaDB)
    SELECT
        id, title,
        givenname, othernames, familyname
    FROM
        books
        LEFT JOIN LATERAL(
            SELECT givenname, othernames, familyname
            FROM authors
            WHERE authors.id=books.authorid
        ) AS a ON true;
-- MSSQL
    SELECT
        id, title,
        givenname, othernames, familyname,
        home
    FROM books
        OUTER APPLY (
            SELECT givenname, othernames, familyname
            FROM authors
            WHERE authors.id=books.authorid
        ) AS a;
-- Oracle: Same as MSSQL without AS
```

You'll get something like

id	Truncate	givenname	othernames	familyname
2078	The Duel	Heinrich	[NULL]	von Kleist
503	Uncle Silas	J.	Sheridan	Le Fanu
2007	North and South	Elizabeth	[NULL]	Gaskell
702	Jane Eyre	Charlotte	[NULL]	Brontë
1530	Robin Hood, The …	Alexandre	[NULL]	Dumas
1759	La Curée	Émile	[NULL]	Zola

~ 1201 rows ~

In this case, you can just as readily use a normal outer join to get the same results:

```
SELECT
    books.id, title,
    givenname, othernames, familyname,
    home
FROM books LEFT JOIN authors ON authors.id=books.authorid;
```

The latter form is definitely simpler (we've left off the table aliases for simplicity and qualified the books.id column out of necessity).

On the other hand, if the subquery is an aggregate query, the lateral join is convenient, since you're going to need a subquery anyway: remember you can't mix aggregate and non-aggregate data in a single SELECT statement.

For example, suppose you want a list of customers with the total sales for each customer. You'll need an aggregate query to get the totals, joined to the customers table. You could do this:

```
-- PostgreSQL, MySQL (not MariaDB)
    SELECT
        id, givenname, familyname, total
    FROM
        customers
        LEFT JOIN LATERAL(
            SELECT sum(total) AS total FROM sales
            WHERE sales.customerid=customers.id
```

```
    ) AS totals ON true;
-- MSSQL, Oracle (no AS)
    SELECT
        id, givenname, familyname, total
    FROM
        customers
        OUTER APPLY(
            SELECT sum(total) AS total FROM sales
            WHERE sales.customerid=customers.id
        ) AS totals;
```

This will give you something like this:

Id	givenname	familyname	total
42	May	Knott	3437.72
459	Rick	Shaw	461.00
597	Ike	Andy	[NULL]
186	Pat	Downe	1536.50
352	Basil	Isk	573.00
576	Pearl	Divers	[NULL]
~ 303 rows ~			

Although there may be alternatives, as you'll need when working with SQLite or MariaDB, the lateral join can sometimes make this sort of query a little more intuitive.

Working with Common Table Expressions

We've seen that you can use subqueries in the FROM clause and that you can even nest them. However, there's an alternative method which makes working with these subqueries more natural.

Common Table Expressions, or CTEs to their friends, are subqueries which generate table results. As a result, you can almost always replace a FROM subquery with a CTE.

We've already used CTEs in previous chapters, so if you get a sense of *déjà vu*, it's OK.

267

Common Table Expressions (CTEs) are a relatively new feature in SQL, but have been around for some time, and are available in almost all modern DBMSs. The notable laggards are MariaDB which added support in version 10.2 (released in 2016) and MySQL which added support in version 8.0 (released in 2018). If you're stuck with an older version of MariaDB or MySQL, maybe you can learn to enjoy nested subqueries.

A CTE is a virtual table defined before it is used. It is similar to a subquery in that it comprises a SELECT statement, but the syntax is different, and it offers a number of advantages.

In their simplest form, CTEs are an alternative form of subquery. Even so, there is an immediate benefit:

- CTEs are more readable and easier to maintain, because they are defined at the beginning rather than in the middle of a query.

Complex subqueries may have one subquery referring to another. This involves nesting subqueries.

- CTEs can reference previous CTEs without the need for nesting.

Both these benefits relate to readability and maintainability. The third benefit is one which is not available for ordinary subqueries.

- CTEs can refer to themselves; thus, they can be recursive.

Syntax

A Common Table Expression is defined as part of the query, before the main part:

```
WITH cte AS (subquery)
SELECT columns FROM cte;
```

The CTE is given a name, though not necessarily cte of course. Thereafter, it is used as a normal table in the main query. You can define multiple CTEs as follows:

```
WITH
    cte AS (subquery),
    another AS (subquery)
SELECT columns FROM ...;
```

Using a CTE to Prepare Calculations

Remember the subquery with the price groups?

```
SELECT price_group, count(*) AS num_books
FROM (
    SELECT
        id, title,
        CASE
            WHEN price<13 THEN 'cheap'
            WHEN price<=17 THEN 'reasonable'
            WHEN price>17 THEN 'expensive'
        END AS price_group
    FROM books
) AS sq     -- Oracle: ( ... ) sq
GROUP BY price_group;
```

That is easily rewritten as a CTE:

```
-- Prepare Data
    WITH sq AS (
        SELECT
            id, title,
            CASE
                WHEN price<13 THEN 'cheap'
                WHEN price<=17 THEN 'reasonable'
                WHEN price>17 THEN 'expensive'
            END AS price_group
        FROM books
    )
-- Use Prepared Data
    SELECT price_group, count(*) AS num_books
    FROM sq
    GROUP BY price_group;
```

It doesn't look much different, but the important part is that you now have your query in two parts: the first part defines the subquery, and the second uses it. It's a much better way of organizing your code.

The subquery has been transferred to a CTE at the beginning of the query. From there on, the main SELECT statement references the CTE as if it were just another table.

The advantage is that the query is written according to the plan: first prepare the data, and then use the data.

MSSQL currently doesn't require a semicolon at the end of a statement, but you should be in the habit of using it anyway.

However, the WITH clause has an alternative meaning at the end of a previous SELECT statement, so it will be misinterpreted if you don't end the previous SELECT statement with the semicolon.

Just use the semicolon at the end of every statement, and all will be fine. *Don't* fall for this nonsense:

```
;WITH (...)
```

Here's another example, which we'll use further in the next few chapters. If you look at the sales table:

```
SELECT * FROM sales;
```

you'll see the following:

id	customerid	total	...	ordered_date
39	28	28.00	...	2022-05-15
40	27	34.00	...	2022-05-16
42	1	58.50	...	2022-05-16
43	26	50.00	...	2022-05-16
45	26	17.50	...	2022-05-16
518	50	13.00	...	[NULL]

~ 5549 rows ~

If you want to summarize the table, such as to get monthly totals, the data is too fine-detailed. Instead, you can prepare the data by formatting the ordered as a year-month value:

```
WITH salesdata AS (
    SELECT
    --   PostgreSQL, Oracle
        to_char(ordered,'YYYY-MM') AS month,
        --   MariaDB/MySQL
            --   date_format(ordered,'%Y-%m') AS month,
        --   MSSQL
            --   format(ordered,'yyyy-MM') AS month,
        --   SQLite
            --   strftime('%Y-%m',ordered) AS month,
        total
    FROM sales
)
SELECT month, sum(total) AS daily_total
FROM salesdata
GROUP BY month
ORDER BY month;
```

You'll now get the following summary:

Month	daily_total
2022-05	6966.50
2022-06	12733.00
2022-07	17314.00
2022-08	19093.00
2022-09	20295.50
2022-10	27797.50
~ 14 rows ~	

In real life, much of what you want to summarize isn't in the right form, but you can prepare it in a CTE to get it ready.

We'll have another look at CTEs in Chapter 9, where we'll see more techniques we can apply.

Summary

In this chapter, we've had a look at using variations on subqueries in a query. We've already seen some subqueries in previous chapters, but here we had a closer look at how they work.

Subqueries can be used in any clause. The results of the subquery must match the context of the clause:

- Subqueries in the SELECT clause or in simple WHERE expressions need to return a single value.

- Subqueries used in an IN() expression need to return a single column.

- Subqueries used in the FROM clause need to return a virtual table.

You can also use subqueries in the ORDER BY clause, though you'd probably want to use the expression in the SELECT clause instead.

You can also use subqueries with the WHERE EXISTS expression or in LATERAL joins.

Subqueries in the FROM clause can be nested, though you would probably want to use a Common Table Expression instead.

Correlated and Non-correlated Subqueries

Subqueries can be correlated or non-correlated:

- A non-correlated subquery is independent of the main query and is evaluated once. The results are then used for the main query.

- A correlated subquery is one which references data in the main query. It is evaluated for every row.

A correlated subquery can be expensive, since it's evaluated multiple times, so there may be more suitable alternatives.

The WHERE EXISTS Expression

The WHERE EXISTS tests whether the subquery returns any rows; if it does, the row in the main query is retained; otherwise, it is filtered out.

Using a non-correlated subquery in WHERE EXISTS will generally return all or none of the rows in the main query. Using a correlated subquery will filter selective rows.

You can generally use an IN( ) expression as an alternative, but there are times when WHERE EXISTS allows a more complex subquery. It also bypasses the NOT IN quirk.

LATERAL JOINS (a.k.a. CROSS APPLY)

A LATERAL JOIN, which, in some DBMSs, is a CROSS APPLY, allows you to add columns to each row in the main query:

- You can use a LATERAL JOIN to add calculated columns, and you can chain them, so you can have multiple calculations and intermediate values.

- You can also use a LATERAL JOIN to add multiple columns from a subquery.

Common Table Expressions

A subquery can be used in a FROM clause, but it can be complex, especially if there's a need for multiple subqueries, which will have to be nested.

A Common Table Expression allows you to define a virtual table before the main query, allowing the main query to work like an ordinary query.

You can also chain multiple CTEs together, which is easier to work with than nesting subqueries.

Coming Up

In Chapter 5, we had a look at aggregating data. Generally, aggregate values can't be mixed with non-aggregate values without throwing a few subqueries into the mix.

Window functions are a group of functions which do the job of applying subqueries to each row. There are two main groups of window functions:

- The aggregate functions can be used to apply an aggregate to each row of a non-aggregate query. They can also be used to accumulate or aggregate in groups.

- The sequencing functions can be used to generate a value based on the position of the row in the dataset. They can be used to indicate the row position or some grouping. They can also be used to fetch values from other rows.

With window functions, you'll be able to generate datasets which combine plain data with more analytical data.

CHAPTER 8

Window Functions

So far, you have seen two main groups of calculations:

- Most calculations have been based on table *columns*: For each row, a value is calculated from one or more columns.

- Aggregate queries are used to summarize *rows*: For the whole table, some or all rows are summarized.

Window functions are a group of functions which add row data as columns. We'll be working with three groups of window functions:

- Aggregate functions: You would normally get aggregates as a separate summary of the table data, but aggregate window functions allow you to include aggregates with each row.

  Among other things, you'll see how this can be used to generate running totals.

- Ranking functions: This will generate a value based on the position of the current row within the dataset.

  Using sequencing functions, you can get the row number, the relative ranking, and even groups such as deciles.

- Value functions: You can get data from rows which precede or follow the current row. You can also get the first and last values in each group.

  This will, for example, get you the difference in values between this and some other row.

In this chapter, we'll look at all of these.

© Mark Simon 2023
M. Simon, *Leveling Up with SQL*, https://doi.org/10.1007/978-1-4842-9685-1_8

Window functions are relatively new to SQL, but most modern DBMSs now support them. Again, the laggards are MariaDB, which introduced them in version 10.2, and MySQL which introduced them in version 8.

Before we get started, some of the samples will be working with the `sales` table. That table includes some NULLs for the ordered date/time. Presumably, those sales never checked out.

We've been pretty forgiving so far and filtered them out from time to time, but the time has come to deal with them. We can delete all of the NULL sales as follows:

```
DELETE FROM sales WHERE ordered IS NULL;
```

You'll notice that there's a foreign key from the `saleitems` table to the `sales` table, which would normally disallow deleting the sales if there are any items attacked. However, if you check the script which generates the sample database, you'll notice the `ON DELETE CASCADE` clause, which will automatically delete the orphaned sale items.

Writing Window Functions

A window function generates a value over a set of rows. The set of rows is called a **window**.

The general syntax for window functions is

```
fn() OVER (PARTITION BY columns |
    ORDER BY columns | frame clause)
```

The important part is the `OVER()` clause which generates the window to be summarized.

There are three main window clauses:

- `PARTITION BY`: This calculates the function for the group defined. It is equivalent to `GROUP BY`.

  The default partition is the whole table.

- ORDER BY: This calculates the function cumulatively, in the order defined. In other words, it generates running totals.

    This order does not need to be the same as the table's ORDER BY clause.

- There is also an optional framing clause. This creates a sliding window within the partition.

    The framing clause requires an ORDER BY window clause. By default, the frame is the rows from the beginning to the current row, but that needs to be qualified when we get to that.

In the following samples, there is normally an ORDER BY clause at the end of the SELECT statement, which is the same as what's in the OVER() clause. This isn't necessary, but it makes the results easier to follow.

Simple Aggregate Windows

As you know, you can't mix aggregate functions in non-aggregate queries. For example, here is an aggregate query which won't work:

```
SELECT
    id, givenname, familyname,
    count(*)
FROM customerdetails;
```

However, this will work:

```
SELECT
    id, givenname, familyname,
    count(*) OVER ()
FROM customerdetails;
```

This gives you something like

id	givenname	familyname	count
42	May	Knott	303
459	Rick	Shaw	303
597	Ike	Andy	303
186	Pat	Downe	303
352	Basil	Isk	303
576	Pearl	Divers	303
~ 303 rows ~			

The OVER() clause changes the aggregate function into a window function. This aggregate function will now be generated for each column. You'll see later that the OVER() clause defines any grouping, known as partitions, the order, and the number of rows to be considered in the aggregate.

For such a simple case, you can get the same result with a subquery:

```
SELECT
    id, givenname, familyname,
    (SELECT count(*) FROM customers)
FROM customerdetails;
```

The window function becomes more interesting when you apply one of the window clauses. For example:

```
SELECT
    id, givenname, familyname,
    count(*) OVER (ORDER BY id)
FROM customerdetails;
```

This will give the running count up to and including the current row, in order of id. The actual table results may or may not be in row order, especially if you include other expressions, so it's better to add that to the end:

```
SELECT
    id, givenname, familyname,
    count(*) OVER (ORDER BY id) AS running_count
FROM customerdetails
ORDER BY id;
```

Now you'll get something like

id	Givenname	familyname	running_count
1	Pierce	Dears	1
2	Arthur	Moore	2
5	Ray	King	3
6	Gene	Poole	4
9	Donna	Worry	5
10	Ned	Duwell	6
~ 303 rows ~			

The running_count column looks very much like a simple row number. We'll see later that it's not necessarily the same if the ORDER BY column isn't unique.

Aggregate Functions

Normally, you can't use aggregate functions in a normal query unless you squeeze them into a subquery. However, they can be repurposed as window functions.

Previously, you saw that you can use the expression count(*) OVER () to give the total number on every row. You can also do something similar with the sum() or avg() functions.

For example, suppose you want to compare sales totals with the overall average:

```
SELECT
    id, ordered, total,
    total-avg(total) OVER () AS difference
FROM sales;
```

You'll get something like

id	Ordered	total	difference
39	2022-05-15 21:12:07.988741	28	-33.783
40	2022-05-16 03:03:16.065969	34	-27.783
42	2022-05-16 10:09:13.674823	58.5	-3.283
43	2022-05-16 15:02:43.285565	50	-11.783
45	2022-05-16 16:48:14.674202	17.5	-44.283
518	[NULL]	13	-48.783

~ 5549 rows ~

In a more complicated example, suppose you want to compare how sales each day compare to the rest of the week.

First, you could extract only the day of the week and total from the sales table. You can use either the day name or the day number for this, but let's use the day number:

```
--  PostgreSQL: Sunday=0
    SELECT
        EXTRACT(dow FROM ordered) AS weekday_number,
        total
    FROM sales;
--  MSSQL: Sunday=1
    SELECT
        datepart(weekday,ordered) AS weekday_number,
        total
    FROM sales;
--  Oracle: Sunday=1
    SELECT
        to_char(ordered,'D')+0 AS weekday_number,
        total
    FROM sales;
--  MariaDB/MySQL: Sunday=1
    SELECT
        dayofweek(ordered) AS weekday_number,
        total
```

```
    FROM sales;
--  SQLite: Sunday=0
    SELECT
        strftime('%w',ordered) AS weekday_number
        total
    FROM sales;
```

You'll see they all have a different way to do it, and they can't even agree on the day number. Fortunately, they all agree on the first day of the week:

weekday_number	Total
0	28
1	34
1	58.5
1	50
1	17.5
0	13
~ 5549 rows ~	

Next, put that into a CTE, so you can aggregate them:

```
WITH
    data AS (
        SELECT
            ... AS weekday,
            total
        FROM sales
    )
--  to be done
;
```

Next, you could summarize the data in another CTE:

```
WITH
    data AS (
        SELECT
```

281

```
        ... AS weekday_number,
          total
      FROM sales
  ),
  summary AS (
      SELECT weekday_number, sum(total) AS total
      FROM data
      GROUP BY weekday_number
  )
-- etc
```

Finally, you can compare the daily totals to the grand totals using a window aggregate:

```
WITH
    data AS (...),
    summary AS (...)
SELECT
    weekday_number, total,
    total/sum(total) OVER()
FROM weekday_number
ORDER BY weekday_number;
```

This will give you a day-by-day summary:

weekday_number	total	?column?
0	48182.22	0.147
1	49304	0.151
2	45156.5	0.138
3	45959.5	0.141
4	47528	0.145
5	42372.5	0.13
6	48415.5	0.148

Note that the expression total/sum(total) OVER() is confusing as the OVER() clause seems a little uninvolved. You might prefer to write it as total/(sum(total) OVER ()) to make it clearer that it is, in fact, a single expression. We'll leave that to your preference, but it isn't normally written that way.

You can finish off by giving the calculation an alias, displaying it as a percentage, and sorting by weekday:

```
WITH
    data AS (...),
    summary AS (...)
SELECT
    weekday, total,
    100*total/sum(total) OVER() AS proportion
FROM summary
;
```

If you want to display the percentage symbol, that's up to the DBMS. You can try one of the following:

```
--  PostgreSQL
    to_char(100*total/sum(total) OVER(),'99.9%')
--  MariaDB/MySQL
    format(100*total/sum(total) OVER(),2) || '%'
--  MSSQL
    format(100*total/sum(total) OVER(),'0.0%')
--  SQLite: aka printf(...)
    select format('%.1f%%',100*total/sum(total) OVER())
--  Oracle
    to_char(100*total/sum(total) OVER(),'99.9') || '%'
```

This looks more convincing:

weekday_number	total	proportion
0	48182.22	14.7%
1	49304	15.1%
2	45156.5	13.8%

(continued)

weekday_number	total	proportion
3	45959.5	14.1%
4	47528	14.5%
5	42372.5	13.0%
6	48415.5	14.8%

We've used OVER() to calculate the grand total for the table. However, we can also use a sliding window, as we'll see in the next section.

Aggregate Window Functions and ORDER BY

Recall our introductory sample where we included an ORDER BY clause in the OVER() clause:

```
SELECT
    id, givenname, familyname,
    count(*) OVER (ORDER BY id) AS running_count
FROM customerdetails
ORDER BY id;
```

In this example, the id, being the primary key, is unique. That will give us a false idea of how this works, so let's look at using the height, which is not unique. We'll also filter out the NULL heights to make it more obvious:

```
SELECT
    id, givenname, familyname,
    height,
    count(*) OVER (ORDER BY height) AS running_count
FROM customerdetails
WHERE height IS NOT NULL
ORDER BY height;
```

You'll see some repeated heights and how they affect the window function:

id	givenname	familyname	Height	running_count
597	Ike	Andy	153	2
283	Ethel	Glycol	153	2
451	Fred	Knott	153.8	3
194	Rod	Fishing	154.3	4
534	Minnie	Bus	156.4	6
352	Basil	Isk	156.4	6

~ 267 rows ~

When using ORDER BY in the OVER clause, it means count the number of rows *up to the current value*. That may or may not be what you wanted.

The Framing Clause

In this example, there's an implied **framing** clause, which defaults to this behavior. If you like, you can make it more specific:

```
count(*) OVER (ORDER BY height
    RANGE BETWEEN UNBOUNDED PRECEDING AND CURRENT ROW)
```

That's quite a mouthful, but that's the way the SQL language is developing: Why say something in two words if you can say it in twenty[1]?

Here, the word RANGE refers to the value of height. For example, in the fifth row earlier, the value is the same as the next row, so count(*) includes both.

The obvious alternative is

```
SELECT
    id, givenname, familyname,
    height,
    count(*) OVER (ORDER BY height
```

---

[1] You'll see this sort of thing in all of the newer features in SQL. You might say that SQL is the new COBOL.
COBOL was (and still is) an early programming language which was supposed to appeal to less mathematical business programmers. It is noted for its verbosity.

285

```
        ROWS BETWEEN UNBOUNDED PRECEDING AND CURRENT ROW) AS running_count
FROM customerdetails
WHERE height IS NOT NULL
ORDER BY height;
```

The subtle change is from RANGE BETWEEN to ROWS BETWEEN. It now counts the number of rows up to the current *row*.

id	givenname	familyname	Height	running_count
597	Ike	Andy	153	1
283	Ethel	Glycol	153	2
451	Fred	Knott	153.8	3
194	Rod	Fishing	154.3	4
534	Minnie	Bus	156.4	5
352	Basil	Isk	156.4	6

~ 267 rows ~

It's a little bit unfair: two customers on the same height are arbitrarily positioned one before the other. We'll see more of this unfairness later.

The framing clause can take the following form:

```
[ROW|RANGE] BETWEEN start AND end
```

As we saw, the difference between ROWS and RANGE is that RANGE includes all the rows which match the current value, while ROWS doesn't.

The start and end expressions, a.k.a. the **frame borders**, can take one of the following forms:

Expression	Meaning
UNBOUND PRECEDING	Beginning
n PRECEDING	Number of rows *before* the current row
CURRENT ROW	
n FOLLOWING	Number of rows *after* the current row
UNBOUND FOLLOWING	End

There's also a short form:

```
ROWS|RANGE start
```

which means between the start and the current row.

Creating a Daily Sales View

Before we proceed, some of our further examples will require some prepared sales data. Although we could do this in a common table expression, it would make sense to prepare a view instead and save ourselves a bit of bother later.

We're going to want daily sales, together with the month of sale. The view would look like this:

```
CREATE VIEW daily_sales AS
SELECT
    ordered_date,

    --   PostgreSQL, Oracle
         to_char(ordered_date,'YYYY-MM') AS ordered_month,
    --   MariaDB/MySQL
         --   date_format(ordered_date,'%Y-%m')
             AS ordered_month,
    --   MSSQL
         --   format(ordered_date,'yyyy-MM') AS ordered_month,
    --   SQLite
         --   strftime('%Y-%m',ordered_date) AS ordered_month,

    sum(total) AS daily_total
FROM sales
WHERE ordered IS NOT NULL
GROUP BY ordered_date;
```

(Don't forget to wrap the statement between GOs for MSSQL.)

We can put that to the test:

```
SELECT * FROM daily_sales ORDER BY ordered_date;
```

You should see something like this:

ordered_date	ordered_month	daily_total
2022-05-04	2022-05	43
2022-05-05	2022-05	150.5
2022-05-06	2022-05	110.5
2022-05-07	2022-05	142
2022-05-08	2022-05	214.5
2022-05-09	2022-05	16.5
~ 389 rows ~		

A Sliding Window

Here's an example of using a sliding window with the framing clause. Suppose we want to generate the daily totals for each day and the week up to the day. We can use

```
SELECT
    ordered_date, daily_total,
    sum(daily_total) OVER(ORDER BY ordered_date
        ROWS 6 PRECEDING) AS week_total,
    sum(daily_total) OVER(ORDER BY ordered_date
        ROWS UNBOUNDED PRECEDING) AS running_total
FROM daily_sales
ORDER BY ordered_date;
```

For both framing clauses, we've used the shorter form, since we want to go up to the current row. We could have left off the framing clause altogether for the running total, but we needed to change from the default RANGE BETWEEN just in case two daily totals were the same.

You'll get something like the following:

ordered_date	daily_total	week_total	running_total
2022-05-04	43	43	43
2022-05-05	150.5	193.5	193.5
2022-05-06	110.5	304	304
2022-05-07	142	446	446
2022-05-08	214.5	660.5	660.5
2022-05-09	16.5	677	677
2022-05-10	160	837	837
2022-05-11	115	909	952
2022-05-12	205	963.5	1157
2022-05-13	164.5	1017.5	1321.5
2022-05-14	46.5	922	1368
2022-05-15	457.5	1165	1825.5

~ 389 rows ~

Note that for the first seven days, the week and running totals are the same, because there are no totals from before then. However, from there on, the running total keeps accumulating while the week total is clamped to the current seven days.

If you look hard enough, you may also see some gaps in the dates. That means that there were no sales on those days and can also mean trouble for interpreting what you mean, since one row is not necessarily one day. We'll address that problem in Chapter 9.

Remember, you're not limited to the count() and sum() functions. For example, you can create sliding averages as well:

```
SELECT
    ordered_date, daily_total,
    sum(daily_total) OVER(ORDER BY ordered_date
        ROWS 6 PRECEDING) AS week_total,
    avg(daily_total) OVER(ORDER BY ordered_dat
        ROWS 6 PRECEDING) AS week_average,
    sum(daily_total) OVER(ORDER BY ordered_date
```

```
        ROWS UNBOUNDED PRECEDING) AS running_total
FROM daily_sales
ORDER BY ordered_date;
```

The week average is the average over the seven days including the current day:

ordered_date	daily_total	week_total	week_average	running_total
2022-05-04	43	43	43	43
2022-05-05	150.5	193.5	96.75	193.5
2022-05-06	110.5	304	101.333	304
2022-05-07	142	446	111.5	446
2022-05-08	214.5	660.5	132.1	660.5
2022-05-09	16.5	677	112.833	677
2022-05-10	160	837	119.571	837
2022-05-11	115	909	129.857	952
2022-05-12	205	963.5	137.643	1157
2022-05-13	164.5	1017.5	145.357	1321.5
2022-05-14	46.5	922	131.714	1368
2022-05-15	457.5	1165	166.429	1825.5
~ 389 rows ~				

You can also select sliding minimums and maximums or averages so far. You'll have to decide which of them is useful for your own purposes.

Window Function Subtotals

Earlier, we created grand totals with an expression like sum(total) OVER(). The OVER() expression is a shorthand for summing over the entire table.

It's also possible to sum (or count, or whatever you like) over groups. You might have thought it would be something like sum(total) OVER (GROUP BY ...), but that's too obvious. Instead, we use the expression (PARTITION BY ...) which means grouping.

The default partition is the whole table. You can partition by anything that can be grouped. For example, suppose you want to get monthly totals with the previous examples, you can use

```
SELECT
    ordered_date, daily_total,
    sum(daily_total) OVER(ORDER BY ordered_date
        ROWS 6 PRECEDING) AS week_total,
    sum(daily_total) OVER(ORDER BY ordered_date
        ROWS UNBOUNDED PRECEDING) AS running_total,
    sum(daily_total) OVER(PARTITION BY ordered_month)
        AS monthly_total
FROM daily_sales
ORDER BY ordered_date;
```

You'll now see something like

ordered_date	daily_total	week_total	running_total	monthly_total
2022-05-04	43	43	43	6966.5
2022-05-05	150.5	193.5	193.5	6966.5
2022-05-06	110.5	304	304	6966.5
2022-05-07	142	446	446	6966.5
2022-05-08	214.5	660.5	660.5	6966.5
2022-05-09	16.5	677	677	6966.5
2022-05-10	160	837	837	6966.5
2022-05-11	115	909	952	6966.5
2022-05-12	205	963.5	1157	6966.5
2022-05-13	164.5	1017.5	1321.5	6966.5
2022-05-14	46.5	922	1368	6966.5
2022-05-15	457.5	1165	1825.5	6966.5

~ 389 rows ~

For every month, you'll get a new total, of course.

Now, here's the tricky part. You can also combine PARTITION BY with ORDER BY:

```
sum(daily_total) OVER(
    PARTITION BY ordered_month
    ORDER BY ordered_date ROWS UNBOUNDED PRECEDING
) AS month_running_total
```

Here's a sample of various possibilities:

```
SELECT
    ordered_date, daily_total,
    sum(daily_total) OVER(ORDER BY ordered_date
        ROWS UNBOUNDED PRECEDING) AS running_total,
    sum(daily_total) OVER(PARTITION BY ordered_month)
        AS month_total,
    sum(daily_total) OVER(ORDER BY ordered_month)
        AS running_month_total,
    sum(daily_total) OVER(PARTITION BY ordered_month
        ORDER BY ordered_date ROWS UNBOUNDED PRECEDING)
        AS month_running_total
FROM daily_sales
ORDER BY ordered_date;
```

You'll see something like this (the column names have been abbreviated to fit in the page):

ordered_date	daily_total	Rt	mt	rmt	mrt
2022-05-04	43	43	6966.5	6966.5	43
2022-05-05	150.5	193.5	6966.5	6966.5	193.5
2022-05-06	110.5	304	6966.5	6966.5	304
2022-05-07	142	446	6966.5	6966.5	446
2022-05-08	214.5	660.5	6966.5	6966.5	660.5
2022-05-09	16.5	677	6966.5	6966.5	677
2022-05-10	160	837	6966.5	6966.5	837
2022-05-11	115	952	6966.5	6966.5	952
2022-05-12	205	1157	6966.5	6966.5	1157

(*continued*)

ordered_date	daily_total	Rt	mt	rmt	mrt
2022-05-13	164.5	1321.5	6966.5	6966.5	1321.5
2022-05-14	46.5	1368	6966.5	6966.5	1368
2022-05-15	457.5	1825.5	6966.5	6966.5	1825.5

~ 389 rows ~

The names may be somewhat confusing, so here's a table of what's going on:

Clause	Name	What's Happening
ORDER BY date ...	running_total	Total so far from the beginning to the current row
PARTITION BY month	month_total	Total for the current group
ORDER BY month	running_month_total	Running total for *each* month
PARTITION BY monthORDER BY date ...	month_running_total	Running total *within* each month

(Again, the column names have been abbreviated to make it all fit.)

Notice how we're using the group column ordered_month both to partition and for a running total. Because its default frame is RANGE ..., it will produce the total for all of the values so far, which effectively is a total for the whole month. This is the sort of thing you can expect if you order by a non-unique row.

The hardest part of it all is thinking of good names for the results.

As summaries, these are all good candidates for saving as a view.

Note, however, that *in SQL Server only*, you cannot include an ORDER BY clause in a view without additional trickery. As a result, you should at least make sure that your SELECT statement includes the columns you want to order by, and then include the ORDER BY clause when using the view.

Alternatively, you can finish the ORDER BY clause with OFFSET 0 ROWS as a workaround.

PARTITION BY Multiple Columns

Given that PARTITION BY generates subtotals, PARTITION BY multiple columns will generate subsubtotals, if that's a real word.

Suppose, for example, that you want to generate a report of sales by state, town, and customer. That data is available, but it's in multiple tables, and you'll need to prepare it first.

First, you'll need to join the customerdetails view (which has the state and town) with the sales. When the time comes, we'll put that in a CTE called customer_sales:

```
-- customer_sales
    SELECT c.id AS customerid, c.state, c.town, total
    FROM customerdetails AS c JOIN sales AS s
        ON c.id=s.customerid
```

We'll then want to summarize the data by grouping by state, town, and customer id. Again, that will go into another CTE:

```
-- totals
    SELECT state, town, customerid, sum(total) AS total
    FROM customer_sales
    GROUP BY state, town, customerid
```

We can put this together and check the results:

```
WITH
    customer_sales AS (
        SELECT c.id AS customerid, c.state, c.town, total
        FROM customerdetails AS c JOIN sales AS s
            ON c.id=s.customerid
    ),
    totals AS (
        SELECT state, town, customerid, sum(total) AS total
        FROM customer_sales
        GROUP BY state, town, customerid
    )
```

```
SELECT state, town, customerid, total AS customer_total
FROM totals
ORDER BY state, customerid;
```

You'll get a result like this:

state	Town	customerid	customer_total
ACT	Kingston	85	2469
ACT	Kingston	112	1387
ACT	Kingston	147	2439.5
ACT	Kingston	355	689.5
ACT	Gordon	489	199
NSW	Reedy Creek	10	3089

~ 269 rows ~

Now for the window functions. First, to get the group total by state, we can use

```
sum(total) OVER(PARTITION BY state) AS state_total
```

To get the group total per town, remember that the town name can appear in more than one state. To use PARTITION BY town would be a mistake, as the town names would be conflated. Instead, we use

```
sum(total) OVER(PARTITION BY state, town) AS town_total
```

Incorporating these two expressions and adding an ORDER BY clause to see it all, we get

```
WITH
    customer_sales AS (
        SELECT c.id AS customerid, c.state, c.town, total
        FROM customerdetails AS c JOIN sales AS s
            ON c.id=s.customerid
    ),
    totals AS (
        SELECT state, town, customerid, sum(total) AS total
```

```
        FROM customer_sales
        GROUP BY state, town, customerid
    )
SELECT
    state, town, customerid, total AS customer_total,
    sum(total) OVER(PARTITION BY state) AS state_total,
    sum(total) OVER(PARTITION BY state, town) AS town_total
FROM totals
ORDER BY state, customerid;
```

The results look like this:

state	Town	customerid	customer_total	state_total	town_total
ACT	Kingston	85	2469	7184	6985
ACT	Kingston	112	1387	7184	6985
ACT	Kingston	147	2439.5	7184	6985
ACT	Kingston	355	689.5	7184	6985
ACT	Gordon	489	199	7184	199
NSW	Reedy Creek	10	3089	106389.22	12655
~ 269 rows ~					

There's an implied hierarchy between a state and a town: a town is part of a state (and, for the time being, a customer is in a town). As a result, the PARTITION BY clause must follow the hierarchy: state,town. You can also use columns which are unrelated, such as the state and year of birth, in which case the columns can go either way.

Ranking Functions

The window functions used so far are basically aggregate functions given a new context. The other group of functions are specific to window functions. Generally, they relate to the position of the current row. Broadly, we can call them ranking functions.

There is one aggregate window function, which we've already seen, which also acts as a ranking function:

```
SELECT
    id, givenname, familyname,
    height,
    count(*) OVER (ORDER BY height
        ROWS UNBOUNDED PRECEDING) AS running_count
FROM customers
WHERE height IS NOT NULL
ORDER BY height;
```

As long as you use the framing clause ROWS UNBOUNDED PRECEDING (shortened from ROWS BETWEEN UNBOUNDED PRECEDING AND CURRENT ROW), the count(*) will count the number of rows up to the current row, which is basically the row number in the result set.

There's a simpler alternative to that:

```
SELECT
    id, givenname, familyname,
    height,
    row_number() OVER (ORDER BY height) AS running_count
FROM customers
WHERE height IS NOT NULL
ORDER BY height;
```

The row_number() function basically generates just that: a number for each row in the result set.

Basic Ranking Functions

There are four main ranking functions:

- row_number(): Calculate the current row number in the current partition in the specified order.

  If two values in the ORDER BY clause are the same, they will still get a different row number; there is no guarantee which one comes first.

- rank(): Give the rank within the result set.

297

If two values in the ORDER BY clause are the same, they will get the same rank. The next different value will *not* get the next rank; it will catch up with the row number above.

- count(*): If you leave the framing clause out and let it default to RANGE, it will behave like rank() with one difference. We'll look at the difference later.

- dense_rank(): This will also give a rank, similar to rank() earlier. However, the next different value *will* get the next rank, so this will gradually fall behind the row number.

If the partition isn't specified (there is no PARTITION BY clause), then the preceding functions apply to the whole table. Otherwise, they will give the position within the group.

The difference between rank() and dense_rank() is that for equal values, rank() will pick up from the next row_number(), while dense_rank() won't.

If the ORDER BY value is not unique

- row_number() is arbitrary.

- rank() gives the rank at the *beginning* of the group.

- count(*) gives the rank at the *end* of the group.

- dense_rank() gives the rank of the group.

If the ORDER BY value is unique, these all give the same results.

We can test this with customer heights, where we know some heights are repeated:

```
SELECT
    id, givenname, familyname,
    height,
    row_number() OVER (ORDER BY height) AS row_number,
    count(*) OVER (ORDER BY height) AS count,
    rank() OVER (ORDER BY height) AS rank,
    dense_rank() OVER (ORDER BY height) AS dense_rank
FROM customers
WHERE height IS NOT NULL
ORDER BY height;
```

You'll get something like this:

id	...	height	row_number	count	rank	dense_rank
597	...	153	1	2	1	1
283	...	153	2	2	1	1
451	...	153.8	3	3	3	2
194	...	154.3	4	4	4	3
534	...	156.4	5	6	5	4
352	...	156.4	6	6	5	4
~ 267 rows ~						

Your actual results may, of course, be different. However, in the preceding example, we can see

- The row_number() is distinct, regardless of the actual value.

- The rank() is the same for equal values. The *next* value matches the row_number().

- The count(*) is also the same for equal values. The *next* value also matches the row_number().

- The rank() is the same as the *first* row_number() for equal values; the count(*) is the same as the *last* row_number() for equal values.

- The dense_rank() is also the same for equal values. The *next* value gets the next rank. By the time you get to the end of the result set, it will be very different to the row number.

With most DBMSs, the ranking functions all require an ORDER BY window clause. That makes sense, since ranking is meaningless without order.

The exceptions include PostgreSQL and SQLite, which will allow an empty window clause:

```
-- PostgreSQL, SQLite
SELECT
    id, givenname, familyname,
    height,
```

```
        row_number() OVER () AS row_number,
        count(*) OVER () AS count,
        rank() OVER () AS rank,
        dense_rank() OVER () AS dense_rank
    FROM customers
    WHERE height IS NOT NULL
    ORDER BY height;
```

However, the results are meaningless. The count(*), rank(), and dense_rank() expressions all give one value for the whole result set, and the row_number() gives row numbers in an arbitrary order.

Ranking with PARTITION BY

By default, ranking functions such as row_number() rank over the entire result set. You can also rank over groups using PARTITION BY:

```
SELECT
    id, ordered_date, total,
    row_number() OVER (PARTITION BY ordered_date) AS row_number
FROM sales
ORDER BY ordered;
```

The results will be something like this:

id	ordered_date	total	row_number
1	2022-05-04	43	1
2	2022-05-05	54.5	1
3	2022-05-05	96	2
6	2022-05-06	18	2
7	2022-05-06	92.5	1
4	2022-05-07	17.5	1

~ 5295 rows ~

The row numbers may not be in the expected order, since the order wasn't specified. To finish the job, we should also include that:

```
SELECT
    id, ordered_date, total,
    row_number() OVER (
        PARTITION BY ordered_date ORDER BY ordered
    ) AS row_number
FROM sales
ORDER BY ordered;
```

The row number is now in the order we would have expected:

id	ordered_date	total	row_number
1	2022-05-04	43	1
2	2022-05-05	54.5	1
3	2022-05-05	96	2
6	2022-05-06	18	1
7	2022-05-06	92.5	2
4	2022-05-07	17.5	1

~ 5295 rows ~

You can use the group row number in a creative way. For example, you might want to show the date for only the first sale for the day. You can show the date selectively using a CASE ... END expression:

```
CASE
    WHEN row_number() OVER
        (PARTITION BY ordered_date ORDER BY ordered)=1
        THEN CAST(ordered_date AS varchar(16))
    ELSE ''
END AS ordered_date,
```

Rearrange and rename a few columns, and you'll get

```
SELECT
    id,
    CASE
        WHEN row_number() OVER
```

```
            (PARTITION BY ordered_date ORDER BY ordered)=1
            THEN CAST(ordered_date AS varchar(16))
        ELSE ''
    END AS ordered_date,
    row_number() OVER (PARTITION BY ordered_date) AS item,
    total
FROM sales
ORDER BY ordered;
```

which gives you a simpler looking result:

id	ordered_date	item	Total
1	2022-05-04	1	43
2	2022-05-05	1	54.5
3		2	96
6	2022-05-06	1	18
7		2	92.5
4	2022-05-07	1	17.5
5		2	63
9		3	61.5
10	2022-05-08	1	67.5
11		2	18.5
8		3	54
13		4	74.5
~ 5295 rows ~			

Of course, you can still include your running totals.

Paging Results

One reason why you might want the overall row number is that you might want to break up your results into pages. For example, suppose you want your results in pages of, say, twenty, and you now want to display page 3 of that.

302

We can start with our `pricelist` view and include the `row_number()` window function:

```
SELECT
    id, title, published, author,
    price, tax, inc,
    row_number() OVER(ORDER BY id) AS row_number
FROM aupricelist;
```

We haven't yet included an `ORDER BY` clause, because there's more to come. Some DBMSs may decide to produce the results in `id` order, but that's not guaranteed, of course.

We can now put this in a CTE and filter on the row number:

```
WITH cte AS (
    SELECT
        id, title, published, author,
        price, tax, inc,
        row_number() OVER(ORDER BY id) AS row_number
    FROM aupricelist
)
SELECT *
FROM cte
WHERE row_number BETWEEN 40 AND 59
ORDER BY id;
```

You'll get something like

id	title	...	price	tax	inc	row_number
98	Camilla	...	12	1.2	13.2	40
102	The Mystery of a Hansom ...	...	14.5	1.45	15.95	41
103	Persian Letters	...	15.5	1.55	17.05	42
104	Sinners in the Hands of ...	...	19.5	1.95	21.45	43
106	Trafalgar	...	16	1.6	17.6	44
109	The Scarlet Letter and S ...	...	19.5	1.95	21.45	45

~ 20 rows ~

303

Oracle has a built-in value called rownum. Sadly, you still need to use it from a CTE or a subquery.

Of course, you don't have to order by the id. You can use the title, or the price, as long as you include it in both the window function and in the ORDER BY clause. And, of course, you can also use DESC.

There is an alternative way to do this. Officially, you can use the OFFSET ... FETCH ... clause:

```
-- PostgreSQL, MSSQL, Oracle
   SELECT
        id, title, published, author,
        price, tax, inc,
        row_number() OVER(ORDER BY id) AS row_number
   ORDER BY id OFFSET 40 ROWS FETCH FIRST 20 ROWS ONLY;
```

This skips over the first 40 rows and fetches the next 20 rows after that. Unofficially, some DBMSs support LIMIT ... OFFSET:

```
-- PostgreSQL (again), MariaDB/MySQL, SQLite
   SELECT
        id, title, published, author,
        price, tax, inc,
        row_number() OVER(ORDER BY id) AS row_number
   ORDER BY id LIMIT 20 OFFSET 40;
```

This is a simpler syntax, but, unfortunately, not the official syntax.

MSSQL also supports the simple SELECT TOP syntax, but it's not so flexible.

Of course, these two alternatives are much simpler than using the window function technique, but there is an advantage with using the window function.

Suppose you're sorting by something non-unique, such as the price. The problem with the normal paging techniques, including the row_number() earlier, is that the page stops strictly at the number of rows (or less if there are no more).

If you decide to keep the prices together, you can instead use something like

```
WITH cte AS (
    SELECT
        id, title, published, author,
        price, tax, inc,
        rank() OVER(ORDER BY price) AS rank
    FROM aupricelist
)
SELECT *
FROM cte
WHERE rank BETWEEN 40 AND 59
ORDER BY price;
```

As long as the groupings aren't too big, it should give you nearly the same results, but with all the books of one price together.

Working with ntile

If you want to split your ordered result set into, say, ten groups, we refer to the groups as **deciles**, from the Latin word for ten. If you want five groups, then they're called **pentiles**, and one hundred groups would be **percentiles**. If you know enough Latin, you can go on to have seven groups or thirteen groups.

Mathematicians have a generic name for any number, called **n**, which is rather catchy once you get used to it. If you separate your ordered data into groups, you create **ntiles**, and the window function is ntile(n), where n is the number of groups.

For example, to create deciles by height in your customers table, you can use

```
SELECT
    id, givenname, familyname, height,
    ntile(10) OVER (order by height) AS decile
FROM customers
WHERE height IS NOT NULL;
```

You'll get something like this:

id	givenname	familyname	height	decile
597	Ike	Andy	153	1
283	Ethel	Glycol	153	1
451	Fred	Knott	153.8	1
194	Rod	Fishing	154.3	1
534	Minnie	Bus	156.4	1
352	Basil	Isk	156.4	1

~ 267 rows ~

Notice that we've filtered out the NULL heights. If we hadn't, then the first or last decile or so will be filled with NULL heights, depending on your DBMS. This creates a group that doesn't really belong, but are included anyway.

That's just one trap with ntile(). There are two traps, one of which might be a deal breaker.

First, note that the preceding result has 267 rows, which doesn't evenly divide by 10. That's OK, but SQL has to work this one out, and you'll find that the first seven groups will have 27 rows, and the rest 26. Of course, your own results may be different, but the idea is the same: the remainder rows will fill in from the front.

The second trap might take some hunting and may not be apparent in your own sample database. If you look hard enough, you may find something like this:

id	givenname	familyname	height	decile
...				
388	Ron	Delay	166.9	3
546	Pat	Ella	167.1	3
106	Jay	Walker	167.1	3
77	Lyn	Seed	167.1	4
403	Will	Knott	167.3	4
314	Jack	Potts	167.4	4
...				

In this sample, you'll see that three customers have the same height (167.1), but one of them didn't fit in the earlier decile, so was pushed into the next. That's more of the unfairness mentioned earlier, as is due to the fact that ntiles are calculated purely on the row number and the value.

If you were, for example, awarding prizes or discounts to customers in certain deciles, it would be unfair to miss out just because the sort order is unpredictable.

This might be a deal breaker, if you rely on the ntile. There is, however, a workaround.

A Workaround for ntile

As we noted, the ntile is based on the row number. If, however, the ntile were based on the rank(), count(), or even dense_rank(), then rows with the same value would end up in the same decile.

In this case, we'll generate twenty groups, called vigintiles. To do that, we'll have to calculate our own groupings. We begin with calculating the size of each group:

```
SELECT count(*)/20.0 AS bin
FROM customers WHERE height IS NOT NULL
```

We'll call this value bin, which is a common statistical name for groups.

We can put that into a CTE and run the following:

```
--  PostgreSQL, MariaDB/MySQL, MSSQL, Oracle
    WITH data AS (
        SELECT count(*)/20.0 AS bin
        FROM customers WHERE height IS NOT NULL
    )
    SELECT
        id, givenname, familyname, height,
        row_number() OVER(ORDER BY height) AS row_number,
        ntile(20) OVER(ORDER BY height) AS vigintile,
        floor((row_number() OVER(ORDER BY height)-1)/bin)+1
            AS row_vitintile,
        floor((rank() OVER(ORDER BY height)-1)/bin)+1
            AS rank_vigintile,
        floor((count(*) OVER(ORDER BY height)-1)/bin)+1
```

```
        AS count_vigintile,
    bin
  FROM customers, data
  WHERE height IS NOT NULL
  ORDER BY height;
```

SQLite doesn't have a `floor()` function, but you can use `cast(... AS int)` instead:

```
cast((row_number() OVER(ORDER BY height)-1)/bin AS int)+1
    AS row_vigintile,
cast((rank() OVER(ORDER BY height)-1)/bin AS int)+1
    AS rank_vigintile,
cast((count(*) OVER(ORDER BY height)-1)/bin AS int)+1
    AS count_vigintile,
```

You'll get the following results:

id	...	height	rn	vig	row_vig	rank_vig	count_vig	bin
597	...	153	1	1	1	1	1	13.35
283	...	153	2	1	1	1	1	13.35
451	...	153.8	3	1	1	1	1	13.35
194	...	154.3	4	1	1	1	1	13.35
534	...	156.4	5	1	1	1	1	13.35
352	...	156.4	6	1	1	1	1	13.35

~ 267 rows ~

Note that the `vigintile` and `row_vigintile` values should be the same; the `row_vigintile` is there to show how the vigintile was calculated from the row number.

More importantly, you'll see that the `rank_vigintile` and `count_vigintile` columns are calculated from the `rank()` and `count(*)` values, and they always put the rows with the same height in the same group. It's up to you to decide which is preferable.

Working with Previous and Next Rows

While working with an ordered result set, we can also get data from previous and next rows. These results are called the `lag` and `lead`, respectively.

The general syntax for the function is

```
lead(column,number) OVER (...)
lag(column,number) OVER (...)
```

Here, as well as the `OVER` clause, we need to supply two values. The `column` value refers to which data in the other row you want. The `number` value refers to how many rows back or forward to get it from. If you want, you can leave it out, in which case it will default to 1.

For example, suppose you want to look at sales for each day, as well as for the previous and next days. You can write

```
SELECT
    ordered_date, daily_total,
    lag(daily_total) OVER (ORDER BY ordered_date)
        AS previous,
    lead(daily_total) OVER (ORDER BY ordered_date)
        AS next
FROM daily_sales
ORDER BY ordered_date;
```

You'll see:

ordered_date	daily_total	previous	next
2022-05-04	43	[NULL]	150.5
2022-05-05	150.5	43	110.5
2022-05-06	110.5	150.5	142
2022-05-07	142	110.5	214.5
2022-05-08	214.5	142	16.5
2022-05-09	16.5	214.5	160

~ 388 rows ~

You'll notice that the previous for the first row is NULL; so is the next for the last row.

You might think that's a bit pointless if you can just move your eyes to look up or down a row. However, you can also incorporate the lag or lead in a calculation. For example, suppose you want to compare sales for each day to a week before. You could use

```
SELECT
    ordered_date, daily_total,
    lag(daily_total,7) OVER (ORDER BY ordered_date)
        AS last_week,
    daily_total
        - lag(daily_total,7) OVER (ORDER BY ordered_date)
        AS difference
FROM daily_sales
ORDER BY ordered_date;
```

This results in

ordered_date	daily_total	last_week	difference
2022-05-04	43	[NULL]	[NULL]
2022-05-05	150.5	[NULL]	[NULL]
2022-05-06	110.5	[NULL]	[NULL]
2022-05-07	142	[NULL]	[NULL]
2022-05-08	214.5	[NULL]	[NULL]
2022-05-09	16.5	[NULL]	[NULL]
2022-05-10	160	[NULL]	[NULL]
2022-05-11	115	43	72
2022-05-12	205	150.5	54.5
2022-05-13	164.5	110.5	54
2022-05-14	46.5	142	-95.5
2022-05-15	457.5	214.5	243

~ 388 rows ~

Here, the expression `lag(total,7)` gets the value for seven rows before. As you'd expect, the first seven rows have `NULL` for the value.

There are two important conditions if you want to use `lag` or `lead` meaningfully:

- There must be only one row for each instance you want to test. For example, you can't have two rows with the same date.

- There must be no gaps. For example, there can't be a missing date.

That's because we're interpreting each row as one day. If you're just working with a sequence or sales regardless of the date, it won't matter.

If you look carefully (and patiently) through the data, you will find that there are a few missing dates. That means that the previous row isn't always "yesterday," and the seven rows previous isn't always "last week." We'll see how to plug these gaps in Chapter 9.

Summary

Window functions are functions which give a row-by-row value based on a "window" or a group of rows.

Window functions include

- Aggregate functions

  These include all of the major nonwindow aggregate functions, such as `count()` and `sum()`.

- Ranking functions and grouping

  These include `row_number()`, `rank()`, and `dense_rank()` to generate a position, as well as `ntile()` to generate ordered groups.

- Functions which fetch data from other rows

  These include `lag()` and `lead()`.

Window Clauses

A window function features an `OVER()` clause:

```
fn() OVER (...)
```

The OVER() clause includes the following:

- ORDER BY to define the row order of the data.

- PARTITION BY to define subgroups of the data.

- A framing clause which determines whether the data is defined by row number or by value. It also determines the start and end rows of the window.

Coming Up

In Chapter 7, we've already discussed how Common Table Expressions work. In fact, we've used them pretty extensively throughout the book.

In the next chapter, we'll have another look at CTEs and examine some of their more sophisticated features. In particular, we'll have a look at the dreaded recursive CTE.

CHAPTER 9

More on Common Table Expressions

You have already made use of CTEs to prepare data for use in aggregates and other operations.

Here, we will take a further look at some of the more powerful features of CTEs.

CTEs As Variables

In Chapter 4, we tested some calculations with a test value:

```
WITH vars AS (
    SELECT ' abcdefghijklmnop ' AS string
    -- FROM dual    -- Oracle
)
SELECT
    string,
    -- sample string functions
FROM vars;
```

Later in this chapter, we'll see a more sophisticated version of this technique when we look at table literals. For now, let's look at how we can use this.

Some DBMSs as well as all programming languages have a concept of **variables**. A variable is a temporary named value. Where the DBMS supports it, you declare a variable name and assign a value which you use in a subsequent step. For example, in MSSQL, you can write this:

```
-- MSSQL
    DECLARE @taxrate decimal(4,2);
```

© Mark Simon 2023
M. Simon, *Leveling Up with SQL*, https://doi.org/10.1007/978-1-4842-9685-1_9

```
SET @taxrate = 12.5;
SELECT
    id, title,
    price, price/@taxrate/100 AS tax
FROM books;
```

To run this, you would need to highlight all of the statements and run in one go.

This chapter won't focus on these variables, but you'll see more on using variables in Chapter 10. Instead, we'll have a look at using a common table expression to do a similar job.

Strictly speaking, what we're going to use is not variables but **constants**, which means that we will set their value once only. However, we can get away with using the looser term "variable," as it's more generic.

There are two main benefits to defining variables:

- You can specify an arbitrary value once, but use it multiple times.

- You move arbitrary values to a preparation section.

In the preceding CTE example, where we're not working with real data, we simply selected from the CTE itself. In more realist examples, we will cross join the CTE with other tables.

Setting Hard-Coded Constants

One simple use for CTE variables is to set an arbitrary value to be used in the main query. For example, suppose we want to generate a simple price list with an arbitrary tax rate.

We can begin with a CTE to contain the tax rate:

```
WITH vars AS (
    SELECT 0.1 AS taxrate
    --  FROM dual    --  Oracle
)
```

We can now combine the CTE with the books table, using a simple cross join:

```
WITH vars AS (
    SELECT 0.1 AS taxrate
    --  FROM dual   --  Oracle
)
SELECT * FROM books, vars;
```

This looks like the following:

id	authorid	Truncate	published	price	taxrate
2078	765	The Duel	1811	12.5	0.1
503	128	Uncle Silas	1864	17	0.1
2007	99	North and South	1854	17.5	0.1
702	547	Jane Eyre	1847	17.5	0.1
1530	28	Robin Hood, The ...	1862	12.5	0.1
1759	17	La Curée	1872	16	0.1

~ 1201 rows ~

A cross join combines every row from one table to every row from another. Since the vars CTE only has one row, the cross join simply has the effect of adding another column to the books table.

SQL has a more modern syntax for a cross join: books CROSS JOIN vars. Here, we'll use the older syntax because it's simpler and more readable.

We can now calculate the price list with tax:

```
WITH vars AS (SELECT 0.1 AS taxrate)
SELECT
    id, title,
    price, price*taxrate AS tax, price*(1+taxrate) AS total
FROM books, vars;
```

This gives us

Id	Title	price	tax	total
2078	The Duel	12.5	1.25	13.75
503	Uncle Silas	17	1.7	18.7
2007	North and South	17.5	1.75	19.25
702	Jane Eyre	17.5	1.75	19.25
1530	Robin Hood, The Prince of Thieves	12.5	1.25	13.75
1759	La Curée	16	1.6	17.6

~ 1201 rows ~

Of course, we could just as readily have used `0.1` instead of the `taxrate` and dispensed with the CTE and the cross join. However, the CTE has the benefit of allowing us to set the tax rate once at the beginning, where it's easy to maintain and can be used multiple times later.

Deriving Constants

The values don't need to be literal values. You can also derive the values from another query. For example, to get the oldest and youngest customers, first set the minimum and maximum dates in variables:

```
-- vars CTE
   SELECT min(dob) AS oldest, max(dob) AS youngest
   FROM customers
```

You can then cross join that with the `customers` table to get the matching customers:

```
WITH vars AS (
   SELECT min(dob) AS oldest, max(dob) AS youngest
   FROM customers
)
SELECT *
FROM customers, vars
WHERE dob IN(oldest, youngest);
```

You should see something like this:

id	givenname	familyname	...	dob	...
92	Nan	Keen	...	1943-05-18	...
228	Cam	Payne	...	2003-01-27	...
577	Sybil	Service	...	2003-01-27	...
392	Daisy	Chain	...	1943-05-18	...

To get the shorter customers, you can set the average height in a variable:

```
WITH vars AS (SELECT avg(height) AS average FROM customers)
SELECT *
FROM customers, vars
WHERE height<average;
```

This is the sort of thing you can't do otherwise, because the average is an aggregate.

Using Aggregates in the CTE

As we've seen many times, you can't mix aggregates with non-aggregate queries. The solution is always to calculate any aggregates you need separately and then incorporate the results in the next query.

Finding the Most Recent Sales per Customer

Suppose, for example, you want to get details about the most recent sale for each customer. To get the most recent sale, you first need a simple aggregate query:

```
SELECT customerid, max(ordered) AS last_order
FROM sales
GROUP BY customerid;
```

You'll get something like this:

Customerid	last_order
550	2023-04-18 09:18:51.933845
272	2023-04-28 09:15:17.85286
70	2023-04-19 14:00:44.880376
190	2023-04-09 10:12:53.416293
539	2023-04-22 16:14:16.173923
314	2023-04-11 03:33:57.825786
~ 269 rows ~	

Here, we have two important pieces of data: the customer id and the date and time of the most recent order. Using this in a subquery, we can join the results with the customers and sales tables to get more details:

```
WITH cte(customerid, last_order) AS (
    SELECT customerid, max(ordered) AS last_order
    FROM sales
    GROUP BY customerid
)
SELECT
    customers.id AS customerid,
    customers.givenname, customers.familyname,
    sales.id AS saleid,
    sales.ordered_date, sales.total
FROM
    sales
    JOIN cte ON sales.customerid=cte.customerid
        AND sales.ordered=cte.last_order
    JOIN customers ON customers.id=cte.customerid
;
```

We'll get something like this:

customer	Givenname	familyname	sale	ordered_date	total
287	Judy	Free	4209	2023-02-04	50.5
26	Bess	Twishes	4542	2023-02-19	11
368	Sharon	Sharalike	4793	2023-03-01	56
282	Howard	Youknow	4939	2023-03-07	39
395	Holly	Day	4953	2023-03-07	75.5
474	Alf	Abet	5092	2023-03-13	94
~ 266 rows ~					

Note that the CTE was used to join the two tables and act as a filter. We don't actually need its results in the output.

Finding Customers with Duplicate Names

In Chapter 2, we saw how to find duplicates using an aggregate query. We did this to find duplicate names, of which there were some, and duplicate phone numbers, of which there were none.

If we were more serious about duplicate customer names, we would want more details about the customers. First, let's find the duplicated names:

```
-- cte
    SELECT familyname, givenname FROM customers
    GROUP BY familyname, givenname HAVING count(*)>1
```

Here, customers are grouped by both names, and the groups are filtered for more than one instance.

Putting that in a CTE, we can join that to the customers table:

```
WITH names AS (
    SELECT familyname, givenname FROM customers
    GROUP BY familyname, givenname HAVING count(*)>1
)
SELECT
    c.id, c.givenname, c.familyname,
```

```
        c.email, c.phone
        --  etc
FROM customers AS c
        JOIN names  ON c.givenname=names.givenname
            AND c.familyname=names.familyname
ORDER BY c.familyname, c.givenname;
```

You'll get something like this:

id	givenname	familyname	email	phone
429	Corey	Ander	corey.ander429@example.net	0355503360
85	Corey	Ander	corey.ander85@example.net	0255501923
174	Paul	Bearer	paul.bearer174@example.com	0370109921
482	Paul	Bearer	paul.bearer482@example.com	0755502522
402	Terry	Bell	terry.bell402@example.com	0755504982
295	Terry	Bell	terry.bell295@example.com	0355509630
~ 16 rows ~				

We've joined the CTE and the customers table using two columns and included their email addresses and phone numbers (if any) so that we can chase them up.

CTE Parameter Names

By default, column names come from the CTE, and you are expected to make sure that all calculations have an alias, as before. If the columns in the CTE don't have an alias, such as when you've calculated something, then (a) you can't refer to the data, and (b) some DBMSs won't let you go ahead.

You can also specify column names with the CTE name as parameters. For example, when we found the first and last dates of birth, we could have put the aliases in the cte expression:

```
WITH vars(oldest, youngest) AS (    -- parameter names
    SELECT min(dob), max(dob)       -- no aliases
    FROM customers
)
```

```
SELECT *
FROM customers, vars
WHERE dob IN(oldest, youngest);
```

For the most part, it's a matter of taste whether you do it this way or add the aliases inside the CTE. If you do include the names, they will override any aliases in the CTE.

One reason you might prefer CTE parameter names is if you think it's more readable, as you have all the names in one place. Later, we'll be writing more complex CTEs which involve multiple CTEs and unions, and it will definitely be easier to follow with parameter names, so you'll be seeing more of that style from here on.

Using Multiple Common Table Expressions

We've seen that, in its simplest form, a CTE can be written as a subquery:

```
SELECT columns
FROM (
    SELECT columns FROM table
) AS sq;
```

A CTE can make this more manageable by putting this subquery at the beginning:

```
WITH cte AS (
    SELECT columns FROM table
)
SELECT columns
FROM cte;
```

That's already an improvement, but where the improvement becomes more obvious is when the subquery also has a subquery:

```
SELECT columns
FROM (
    SELECT columns FROM (
        SELECT columns FROM table
    ) AS sq1
) AS sq2;
```

That's called nesting subqueries, and it can become a nightmare if things get too complex.

Thankfully, CTEs work much more simply:

```
WITH
    sq1 AS (SELECT columns FROM table),
    sq2 AS (SELECT columns FROM sql1)
SELECT columns FROM sq2;
```

You can have multiple CTEs chained this way, as long as you remember to separate them with a comma. As you see in this example, each subquery can refer to a previous one in the chain.

We'll build this up a little more later, and we'll see that additional CTEs don't necessarily have to refer to the previous ones.

Summarizing Duplicate Names with Multiple CTEs

When we produced our list of duplicated names, we had one row for each instance of the name. In Chapter 7, we produced a more consolidated list, but without the benefit of CTEs.

Here, we'll reproduce the consolidated list, but using CTEs to make it much more workable.

We'll start off with the previous query for duplicated names:

```
WITH names AS (
    SELECT familyname, givenname FROM customers
    GROUP BY familyname, givenname HAVING count(*)>1
)
SELECT
    c.id, c.givenname, c.familyname,
    c.email, c.phone
FROM customers AS c
    JOIN names  ON c.givenname=names.givenname
        AND c.familyname=names.familyname
ORDER BY c.familyname, c.givenname;
```

This time, we'll put the results into a second CTE:

```
WITH
    names AS (
        SELECT familyname, givenname FROM customers
        GROUP BY familyname, givenname HAVING count(*)>1
    ),
    duplicates(givenname, familyname, info) AS (
        SELECT
            c.givenname, c.familyname,
            cast(c.id AS varchar(5)) || ': ' || c.email
                --  MSSQL: Use +
        FROM customers AS c      --  Oracle: No AS
            JOIN names ON c.givenname=names.givenname
                AND c.familyname=names.familyname
    )
SELECT * from duplicates
ORDER by familyname, giv1enname;
```

Note

- The layout has changed to make multiple CTEs easier to follow.

- The duplicates CTE has the parameter names for simplicity.
  There's no need to do that with the names CTE, as there are
  no calculated values; however, you may want to do that for
  consistency.

- Instead of listing the id separately, we've cast it to a string and
  concatenated it to the email address. This is to get ready for what
  follows.

- For simplicity, we've ignored the phone number, since it may be
  missing.

You can see what you get so far:

givenname	familyname	info
Corey	Ander	429: corey.ander429@example.net
Corey	Ander	85: corey.ander85@example.net
Paul	Bearer	174: paul.bearer174@example.com
Paul	Bearer	482: paul.bearer482@example.com
Terry	Bell	402: terry.bell402@example.com
Terry	Bell	295: terry.bell295@example.com
~ 16 rows ~		

The next step is to consolidate them by combining the info column values:

```
WITH
    names AS ( ),
    duplicates(givenname, familyname, info) AS ( )
SELECT
    givenname, familyname, count(*),
--  PostgreSQL, MSSQL
    string_agg(info,', ') AS info
--  MySQL/MariaDB
    --  group_concat(info SEPARATOR ', ') AS info
--  SQLite
    --  group_concat(info,', ') AS info
--  Oracle
    --  listagg(info,', ') AS info
FROM duplicates
GROUP BY familyname, givenname
ORDER by familyname, givenname;
```

The consolidated list looks like this:

givenname	familyname	count	info
Corey	Ander	2	429: corey.ander ..., 85: corey.ander8 ...
Paul	Bearer	2	174: paul.bearer ..., 482: paul.bearer ...
Terry	Bell	2	402: terry.bell4 ..., 295: terry.bell2 ...
Mary	Christmas	2	465: mary.christ ..., 594: mary.christ ...
Ida	Dunnit	2	504: ida.dunnit5 ..., 90: ida.dunnit90 ...
Judy	Free	2	93: judy.free93@ ..., 287: judy.free28 ...
Annie	Mate	2	505: annie.mate5 ..., 357: annie.mate3 ...
Ken	Tuckey	2	98: ken.tuckey98 ..., 488: ken.tuckey4 ...

We'll see more examples of multiple CTEs in the following sections.

Recursive CTEs

As you've seen, a feature of using CTEs is that one CTE can refer to a previous CTE. Another feature is that a CTE can refer to itself.

Anything which refers to itself is said to be **recursive**. If you're a programmer, recursive functions are functions which call themselves and are very risky if not handled properly. Similarly, a recursive CTE can be very risky if you're not careful.

A recursive CTE takes one of two forms, depending on your DBMS:

```
-- PostgreSQL, MariaDB/MySQL, SQLite
   WITH RECURSIVE cte AS (
       -- Anchor
          SELECT ...
       UNION
       -- Recursive Member
          SELECT ... FROM cte WHERE ...
   )
-- MSSQL, Oracle
   WITH cte AS (
       -- Anchor
          SELECT ...
```

```
        UNION ALL
        -- Recursive Member
            SELECT ... FROM cte WHERE ...
    )
```

As you see, PostgreSQL, MariaDB/MySQL, and SQLite use the RECURSIVE keyword. MSSQL and Oracle don't, but require a UNION ALL instead of a simple UNION.

In both cases, you'll see that the recursive CTE has two parts:

- The **anchor** defines the starting point or the first member.

  In simple cases, there will be one value, but in other queries there may be more than one.

- The **recursive member** defines data based on what is inherited from the previous iteration of the CTE. That is, it defines the next member.

  Again, if there's more than one anchor member, then there will be multiple recursive members.

Note that the recursive CTE must define when it's going to end or, more correctly, when it can continue. Typically, that's with a WHERE clause, as you've seen earlier, but can use any other method, such as a join.

A simple example of a recursive CTE is one which generates a simple sequence. For example:

```
-- PostgreSQL, MariaDB/MySQL, SQLite
    WITH RECURSIVE cte(n) AS (
        -- Anchor
            SELECT 1
        UNION
        -- Recursive Member
            SELECT n+1 FROM cte WHERE n<10
    )
    SELECT * FROM cte;
-- MSSQL, Oracle
    WITH cte(n) AS (
        -- Anchor
            SELECT 1     -- Oracle: FROM dual
```

```
      UNION ALL
      --  Recursive Member
          SELECT n+1 FROM cte WHERE n<10
 )
 SELECT * FROM cte;
```

The CTE includes a parameter for convenience (cte(n)). Otherwise, you can put the alias in the SELECT statement.

The single anchor value, in this case, is the number 1. The recursive (next) value is n+1, so long as n<10. After that, it stops, and you end up with

N
1
2
3
...
8
9
10

—a sequence of numbers from one to ten.

Recursive CTEs are the closest thing you'll get in standard SQL to iterations or looping.[1]

Two common uses of recursive CTEs are

- Generate a sequence

- Traverse a hierarchy

We'll also use a recursive CTE to split a string into smaller parts, just to show you a little creativity can be added to your queries.

---

[1] Some SQLs, but not all, include additional structures such as DO ... WHILE in an SQL script. They're not really a standard part of the SQL language, but can be used in situations where you're desperate to do something iteratively.

Generating a Sequence

We've already seen how to generate a sequence of numbers:

```
WITH cte AS (
    -- Anchor
        SELECT 0 AS n
    UNION ALL
    -- Recursive
        SELECT n+1 FROM cte WHERE n<100
)
SELECT * FROM cte;
```

The thing to remember is that the recursive member has a WHERE clause to limit the sequence. Without that, the recursive query would try to run forever, and as you know, nothing lasts forever.

MSSQL has a built-in safety limit of 100 recursions, which we'll have to circumvent later:

```
-- MSSQL
    WITH cte (

    )
    SELECT ... FROM cte OPTION(MAXRECURSION ...);
```

The others don't, but for PostgreSQL, MariaDB, and MySQL, you can readily set a time limit:

```
-- PostgreSQL
    SET statement_timeout TO '5s';
-- MariaDB
    SET MAX_STATEMENT_TIME=1;        -- seconds
-- MySQL
    SET MAX_EXECUTION_TIME=1000;    -- milliseconds
```

If you're sure about your recursion terminating properly, you don't need to worry about this. In MSSQL, you will, however, need to increase or disable the recursion limit for some queries.

However, it won't hurt to include a simple number sequence in what follows just to be safe.

One case where a sequence can be useful is to get a sequence of dates. This will simply define a start date and add one day in the recursive member.

The CTE starts simply enough:

```
-- PostgreSQL, MariaDB / MySQL
   WITH RECURSIVE dates(d, n) AS (
       SELECT date'2023-01-01', 1
   )
   SELECT * FROM dates;
-- MSSQL
   WITH dates(d, n) AS (
       SELECT cast('2023-01-01' as date), 1
   )
   SELECT * FROM dates;
-- Oracle
   WITH dates(d, n) AS (
       SELECT date '2023-01-01', 1 FROM dual
   )
   SELECT * FROM dates;
-- SQLite
   WITH RECURSIVE dates(d, n) AS (
       SELECT '2023-01-01', 1
   )
   SELECT * FROM dates;
```

Note that the first value, d, has been cast to a date, with the exception of SQLite, which doesn't have a date type. The n set to 1 is added as a sequence number, but is really unnecessary. It's added here to illustrate how you can use it to stop overrunning your CTE.

The recursive part is also easy enough, but adding one day varies between DBMSs:

```
-- PostgreSQL
   WITH RECURSIVE dates(d, n) AS (
       SELECT date'2023-01-01', 1
       UNION
       SELECT d+1, n+1 FROM dates
       WHERE d<'2023-05-01' AND n<10000
```

```
    )
    SELECT * FROM dates;
-- MariaDB / MySQL
    WITH RECURSIVE dates(d, n) AS (
        SELECT date'2023-01-01', 1
        UNION
        SELECT date_add(d, interval 1 day), n+1 FROM dates
        WHERE d<'2023-05-01' AND n<10000
    )
    SELECT * FROM dates;
-- MSSQL
    WITH dates(d, n) AS (
        SELECT cast('2023-01-01' as date), 1
        UNION ALL
        SELECT dateadd(day,1,d), n+1 FROM dates
        WHERE d<'2023-05-01' AND n<10000
    )
    SELECT * FROM dates;
-- SQLite
    WITH RECURSIVE dates(d, n) AS (
        SELECT '2023-01-01', 1
        UNION
        SELECT strftime('%Y-%m-%d',d,'+1 day'), n+1 FROM dates
        WHERE d<'2023-05-01' AND n<10000
    )
    SELECT * FROM dates;
-- Oracle
    WITH dates(d, n) AS (
        SELECT date '2023-01-01', 1 FROM dual
        UNION ALL
        SELECT d+1, n+1 FROM dates
        WHERE d<date'2023-05-01' AND n<10000
    )
    SELECT * FROM dates;
```

You'll see a series of dates (and numbers):

D	n
2023-01-01	1
2023-01-02	2
2023-01-03	3
2023-01-04	4
2023-01-05	5
2023-01-06	6
~ 121 rows ~	

You'll notice that for MSSQL, we've added `OPTION (MAXRECURSION 0)`, which basically disables the recursion limit.

Note also the `AND n<10000` in the `WHERE` clause. That number is pretty big, and it amounts to over 27 years, but it's not infinite. If you make an error in when to stop the CTE, that expression should limit the recursions.

You might wonder why you would want a sequence of dates between `2023-01-01` and `2023-05-01`, the answer would be "why not?", which isn't very convincing. However, we're going to use this technique to overcome a problem mentioned in Chapter 8: some of the dates will be missing from our summary.

Joining a Sequence CTE to Get Missing Values

You can `JOIN` a recursive CTE which generates a sequence with another table or CTE which has gaps in the sequence to fill in the missing values.

For example, to get the number of customers born per year, it is possible that some years will be missing, but you would like to include the missing year anyway.

First, get a sequence of years:

```
-- PostgreSQL, MariaDB/MySQL, SQLite
WITH RECURSIVE
    allyears(year) AS (
        SELECT 1940
        UNION
```

```
            SELECT year+1 FROM allyears WHERE year<2010
        )

-- MSSQL, Oracle
WITH
    allyears(year) AS (
        SELECT 1940
        UNION ALL
        SELECT year+1 FROM allyears WHERE year<2010
    )
```

Next, get the customer (id) and the year of birth of the customers:

```
-- PostgreSQL, MariaDB/MySQL, Oracle
    yobs(yob) AS (
        SELECT id, EXTRACT(year FROM dob)
        FROM customers WHERE dob IS NOT NULL
    )
-- MSSQL
    yobs(yob) AS (
        SELECT id, year(dob)
        FROM customers WHERE dob IS NOT NULL
    )
-- SQLite
    yobs(yob) AS (
        SELECT id, strftime('%Y',dob)
        FROM customers WHERE dob IS NOT NULL
    )
```

Finally, JOIN them and get the aggregate:

```
WITH RECURSIVE  --  MSSQL, Oracle: no RECURSIVE
    allyears(year) AS ( ),
    yobs AS ( )
SELECT allyears.year, count(*) AS nums
FROM allyears LEFT JOIN yobs ON allyears.year=yobs.yob
GROUP BY allyears.year
ORDER BY allyears.year;
```

You'll need the LEFT JOIN to include all of the sequence of years even if it doesn't match a customer year; after all, that's why it's there.

year	nums
1940	1
1941	1
1942	1
1943	1
1944	1
1945	1
~ 71 rows ~	

We'll do the same sort of thing for sales data.

Daily Comparison Including Missing Days

The same can be applied to missing dates. In Chapter 8, we generated a summary of sales per day. We then created a view with daily sales, such as were available. We can then select from the view:

```
SELECT *
FROM daily_sales
ORDER BY ordered_date;
```

You get something like this:

ordered_date	ordered_month	daily_total
2022-04-08	2022-04	97.5
2022-04-09	2022-04	96
2022-04-10	2022-04	191
2022-04-11	2022-04	201.5
2022-04-12	2022-04	91
2022-04-13	2022-04	160
~ 385 rows ~		

However, if you look hard enough, you'll find some dates missing. We're about to fill them in.

For this, we'll need the following:

- The daily_sales view

- A CTE with the first and last dates of the daily sales

- A sequence of dates

You already know how to generate a sequence of dates. This time, instead of starting and stopping on arbitrary dates, we'll start and stop on the first and last dates of the daily_sales view. We can put those values in a CTE for reference:

```
WITH
    vars(first_date, last_date) AS (
        SELECT min(ordered_date), max(ordered_date)
        FROM daily_sales
    )
```

We can now use these values to generate our sequence of dates:

```
-- PostgreSQL
    WITH RECURSIVE
        vars(first_date, last_date) AS ( ),
        dates(d) AS (
            SELECT first_date FROM vars
            UNION
            SELECT d+1 FROM vars, dates WHERE d<last_date
        )
-- MariaDB / MySQL
    WITH RECURSIVE
        vars(first_date, last_date) AS ( ),
        dates(d) AS (
            SELECT first_date FROM vars
            UNION
            SELECT date_add(d, interval 1 day)
            FROM vars, dates WHERE d<last_date
        )
```

```
-- MSSQL
   WITH
       vars(first_date, last_date) AS ( ),
       dates(d) AS (
           SELECT first_date FROM vars
           UNION ALL
           SELECT dateadd(day,1,d)
           FROM vars, dates WHERE d<last_date
       )
-- SQLite
   WITH RECURSIVE
       vars(first_date, last_date) AS ( ),
       dates(d) AS (
           SELECT first_date FROM vars
           UNION
           SELECT strftime('%Y-%m-%d',d,'+1 day')
           FROM vars, dates WHERE d<last_date
       )
-- Oracle
   WITH
       vars(first_date, last_date) AS ( ),
       dates(d) AS (
           SELECT first_date FROM vars
           UNION ALL
           SELECT d+1 FROM vars, dates WHERE d<last_date
       )
```

For those DBMSs which use the keyword RECURSIVE, you use it once at the
beginning, even if some of the CTEs aren't recursive.

Notice that we've cross-joined the vars and dates, which is the usual technique of
applying variables to another table. We could have written CROSS JOIN, but it's not worth
the effort.

We can now complete our query using a LEFT JOIN to get all of the sequence of dates:

```
WITH RECURSIVE  --  MSSQL, Oracle: no RECURSIVE
    vars(first_date, last_date) AS (
        -- etc
    ),
    dates(d) AS (
        -- etc
    )
SELECT d AS ordered_date, daily_sales.daily_total
FROM dates LEFT JOIN daily_sales ON dates.d=daily_sales.ordered_date
ORDER BY dates.d;
```

We'll now see the following:

ordered_date	daily_total
2022-04-08	97.5
2022-04-09	96
2022-04-10	191
2022-04-11	201.5
2022-04-12	91
2022-04-13	160
~ 387 rows ~	

Notice that we've selected dates.d AS ordered_date in favor of the ordered_date from the daily_sales view. That's because the latter has some missing dates, which is why we went to this trouble in the first place.

Of course, generating a simple sequence isn't the only use for a recursive CTE.

Traversing a Hierarchy

Another use case for a recursive CTE is to traverse a hierarchy. The hierarchy we're going to look at is in the employees table:

```
SELECT * FROM employees;
```

Of course, in a real employees table, there would be more details; we've only included enough here to make the point.

In particular, you'll see that in the employees table, there is a supervisorid column which is a foreign key to the same table:

employees.supervisorid ➤ employees.id

A more naive approach would be either to include the supervisor's name, which is wrong for the same reasons we don't include the author's name with the books table, or to reference another table of supervisors, which is wrong for a different, more subtle reason.

With books and authors, the point is that an author is not the same as a book. In a well-designed database, each table has only one type of member. That's not the case with employees and supervisors. Put simply, the supervisor is another employee.

We're going to traverse the employees table to get a list of employees and their supervisors.

Getting a Single-Level Hierarchy

Without a recursive CTE, you can get employees' supervisors with a simple OUTER JOIN to the same table. This is often referred to as a **self-join**:

```
SELECT
    e.id AS eid,
    e.givenname, e.familyname,
    s.id AS sid,
    s.givenname||' '||s.familyname AS supervisor
    -- s.givenname+' '+s.familyname AS supervisor  --  MSSQL
FROM employees AS e LEFT JOIN employees AS s
    ON e.supervisorid=s.id                 --  Oracle: No AS
ORDER BY e.id;
```

You'll get something like this:

eid	givenname	familyname	sid	supervisor
1	Marmaduke	Mayhem	10	Beryl Bubbles
2	Osric	Pureheart	12	Mildred Thisenthat
3	Rubin	Croucher	[NULL]	[NULL]
4	Gladys	Raggs	29	Fred Nurke
5	Cynthia	Hyphen-Smythe	12	Mildred Thisenthat
6	Sebastian	Trefether	5	Cynthia Hyphen-Smythe

~ 34 rows ~

The trick is, when joining a table to itself, you need to give the table two different aliases to qualify the join.

Multilevel Hierarchy Using Recursive CTE

What we really want is not just the immediate supervisor but a hierarchical list of all supervisors for each employee. This will, of course, require a recursive CTE.

The anchor member will be the employees who have no supervisors, presumably those at the top of the hierarchy:

```
WITH RECURSIVE      -- MSSQL, Oracle: No RECURSIVE
    cte(id, givenname, familyname, supervisorid,
        supervisors, n) AS (
        -- anchor
        SELECT
            id, givenname, familyname, supervisorid, '', 1
        FROM employees WHERE supervisorid IS NULL
    )
```

The columns include some of the raw details, as well as a string of supervisors. Obviously, for the anchor member, the supervisor id will be NULL, and the string is empty. You'll also notice a sequence number, starting at 1. That's for a trick we'll resort to later on.

There will be more than one row for the anchor. That's all right and will still work the same way. There'll just be more than one sequence going.

The recursive member will be the employees with supervisors (i.e., the rest) with a growing list of their supervisors:

```
--  Not MSSQL or MariaDB/MySQL Yet!
WITH RECURSIVE  --  MSSQL, Oracle: No RECURSIVE
    cte(id, givenname, familyname, supervisorid,
        supervisors, n) AS (
        --  anchor
        UNION ALL
        --  recursive: others (supervisorid NOT NULL)
        SELECT
            e.id, e.givenname, e.familyname, e.supervisorid,
            cte.givenname||' '||cte.familyname||' < '||
                cte.supervisors, n+1
        FROM cte JOIN employees AS e ON cte.id=e.supervisorid
        --  Oracle: no AS
    )
SELECT * FROM cte
ORDER BY id;
```

The join is similar to the self-join earlier. The current employee is referred to in the e table alias, and this aliased table is joined to the CTE, which will be the supervisor. The raw data will be from the aliased table, while the supervisor's details will be concatenated as the new supervisors parameter.

Normally, you'd want to limit the recursion with a WHERE clause. For this one, the join will do the job, as it will stop when there are no more to be joined.

The magic is in the expression for the supervisors string. In the recursive member, the CTE represents inherited values.

id	givenname	familyname	sid	supervisors	n
1	Marmaduke	Mayhem	10	Beryl Bubbles < Mildred Thisenth ... <	3
2	Osric	Pureheart	12	Mildred Thisenth ... <	2
3	Rubin	Croucher	[NULL]	[NULL]	1
4	Gladys	Raggs	29	Fred Nurke < Murgatroyd Murdo ... < Rubin Croucher <	4
5	Cynthia	Hyphen-Smythe	12	Mildred Thisenth ... <	2
6	Sebastian	Trefether	5	Cynthia Hyphen-S ... < Mildred Thisenth ... <	3

~ 34 rows ~

This will work in most DBMSs, but not yet in MSSQL or in MariaDB/MySQL. However, it will *nearly* work.

In the case of MariaDB/MySQL, the ' ' in the anchor causes it to jump to the conclusion that the string will be zero characters long, so the supervisors column will be empty.

You will need to cast your empty string in the anchor to a longer one:

```
SELECT
    ..., cast('' AS char(255)), 1
FROM employees WHERE supervisorid IS NULL
```

A long-standing complaint of MySQL is that you don't cast to a varchar, but to a char, which unlike a normal char isn't a fixed length, so it's really a varchar anyway. Nobody knows why. In MariaDB, they allow varchar. The length of 255 should be enough.

It gets worse with MSSQL. Naturally, the columns in the anchor and recursive member should match in data type, but in MSSQL, this match needs to be very exact. The fact that they are both strings is not enough. The strings will need to be of the same type and size. Concatenating the strings produces a longer string, and MSSQL will decide that the longer strings aren't data compatible with the other strings.

In this case, you'll need to cast *both* expressions to the same:

```
SELECT
    ..., cast('' AS nvarchar(255)), ...
FROM employees WHERE supervisorid IS NULL
UNION ALL
SELECT
    ...,
    cast(cte.givenname+' '+cte.familyname
        +' < '+cte.supervisors as nvarchar(255)), ...
FROM cte JOIN employees AS e ON cte.id=e.supervisorid
```

With those changes, the query should work.

Cleaning the Tail End of the List

You'll notice that, with the exception of the empty supervisors strings, there's a trailing <. That's because that character is always added in the recursive member.

If you look at the n column, you'll see that it represents a level number. The character should only be added when there's already another supervisor—that is to say, when we're now adding a second or subsequent supervisor. That means when n has reached two or higher.

We can change that part of the expression by using a CASE ... END expression:

```
-- Others
   cte.givenname||' '||cte.familyname
       || CASE WHEN n>1 THEN ' < ' ELSE '' END
       || cte.supervisors
-- MSSQL
   cte.givenname+' '+cte.familyname
       + CASE WHEN n>1 THEN ' < ' ELSE '' END
       + cte.supervisors
```

This will now produce a cleaner result:

id	givenname	familyname	Sid	supervisors	n
1	Marmaduke	Mayhem	10	Beryl Bubbles < Mildred Thisenth ...	3
2	Osric	Pureheart	12	Mildred Thisenth ...	2
3	Rubin	Croucher	[NULL]	[NULL]	1
4	Gladys	Raggs	29	Fred Nurke < Murgatroyd Murdo ... < Rubin Croucher	4
5	Cynthia	Hyphen-Smythe	12	Mildred Thisenth ...	2
6	Sebastian	Trefether	5	Cynthia Hyphen-S ... < Mildred Thisenth ...	3
~ 34 rows ~					

Of course, we don't need the n column at the end anymore.

Working with Table Literals

At some point, you might want to work with a set of values which haven't been saved anywhere—something in a virtual table which stays around long enough for you to process the data and then discreetly vanishes when you've finished.

In principle, SQL does that all the time when you insert literal values into a table. For example, a statement like

```
INSERT INTO table(columns)
VALUES ( ... ), ( ... ), ( ... );
```

inserts from a virtual table, generated by the VALUES clause. That also means that, in principle, you should be able to use VALUES ... as a virtual table without actually inserting anything. Unfortunately, it's not quite so straightforward.

A **table literal** is an expression which results in a collection of rows and columns—a virtual table. If things go according to plan, it could look like this:

```
VALUES ('a','apple'), ('b','banana'), ('c','cherry')
```

Not all DBMSs see it that way. Some DBMSs do allow just such an expression, but others have something a little more complicated.

A little later, we'll want to work with a virtual table to experiment with, so the first step will be to put this into a CTE. Using the standard notation, you can use

```
-- PostgreSQL, MariaDB (not MySQL), SQLite
   WITH cte(id,value) AS (
       VALUES ('a','apple'), ('b','banana'), ('c','cherry')
   )
   SELECT * FROM cte;
```

You'll see the following:

id	value
a	apple
b	banana
c	cherry

Note that we've included the column names in the CTE name.

For the other DBMSs, there are various alternatives:

```
-- MSSQL
   WITH cte(id,value) AS (
       SELECT * FROM
       (VALUES ('a','apple'), ('b','banana'),
           ('c','cherry')) AS sq(a,b)
   )
   SELECT * FROM cte;
-- MySQL (not MariaDB)
   WITH cte(id,value) AS (
       VALUES ROW('a','apple'), ROW('b','banana'),
           ROW('c','cherry')
   )
   SELECT * FROM cte;
-- Oracle
   WITH cte(id,value) AS (
       SELECT 'a','apple' FROM dual
```

```
     UNION ALL SELECT 'b','banana' FROM dual
     UNION ALL SELECT 'c','cherry' FROM dual
  )
  SELECT * FROM cte;
```

As you see, the prize for the most awkward version goes to Oracle, which doesn't yet support a proper table literal. Apparently, that's coming soon.

MSSQL does support a table literal, but, for some unknown reason, it has to be inside a subquery, complete with a dummy subquery name and dummy column names.

MySQL also supports a table literal, but requires each row inside a ROW() constructor, because MySQL has a non-standard values() function which conflicts with using it simply as a table literal. This is one of the cases where MariaDB and MySQL are not the same.

Using a Table Literal for Testing

One reason you might want throwaway values is if you're testing something, and you haven't the energy to put the test value in a real or temporary value.

For example, suppose you want to test calculating the difference between dates, so for finding an age. You could do something like this:

```
WITH dates(dob,today) AS (
    --  list of dob and today values
)
SELECT
    --  today - dob AS age
FROM dates;
```

The actual code is commented out, because the DBMSs all have their own ways. It gets further complicated because of the date literals.

We're going to try this with the following series of dates:

dob	today
1940-07-07	2023-01-01
1943-02-25	2023-01-01
1942-06-18	2023-01-01
1940-10-09	2023-01-01

(continued)

dob	today
1940-07-07	2022-12-31
1943-02-25	2022-12-31
1942-06-18	2022-12-31
1940-10-09	2022-12-31
1940-07-07	2023-07-07
1943-02-25	2023-02-25
1942-06-18	2023-06-18
1940-10-09	2023-10-09

(If you recognize the dates of birth, don't let on.)

First, you'll need to set up your dates CTE. This is complicated by the fact that in SQL a date literal is in single quotes. However, without context, SQL will regard single quote literals as strings, which won't work with date calculations. The exception is SQLite, which only works with date strings anyway.

The dates CTE would look like this:

```
-- PostgreSQL, MariaDB (not MySQL)
   WITH dates(dob, today) AS (
       VALUES
           (date'1940-07-07',date'2023-01-01'),
           ('1943-02-25','2023-01-01'),
           ('1942-06-18','2023-01-01')
           -- etc
   )
-- MySQL (not MariaDB)
   WITH dates(dob, today) AS (
       VALUES
           row(date'1940-07-07',date'2023-01-01'),
           row('1943-02-25','2023-01-01'),
           row('1942-06-18','2023-01-01')
           -- etc
   )
-- MSSQL
```

```
    WITH dates(dob, today) AS (
        SELECT * FROM (VALUES
            (cast('1940-07-07' as date),
                cast('2023-01-01' as date)),
            ('1943-02-25','2023-01-01'),
            ('1942-06-18','2023-01-01')
            -- etc
        ) AS sq(a,b)
    )
-- SQLite
    WITH dates(dob, today) AS (
        VALUES
            ('1940-07-07','2023-01-01'),
            ('1943-02-25','2023-01-01'),
            ('1942-06-18','2023-01-01')
            -- etc
    )

-- Oracle
    WITH dates(dob, today) AS (
        SELECT date'1940-07-07',date'2023-01-01' FROM dual
        UNION ALL SELECT date'1943-02-25',date'2023-01-01'
            FROM dual
        UNION ALL SELECT date'1942-06-18',date'2023-01-01'
            FROM dual
        -- etc
    )
```

Note

- For PostgreSQL, MariaDB/MySQL, and Oracle, you can use the simple expression date'...' to interpret a literal as a date.

- For PostgreSQL, MariaDB/MySQL, and MSSQL, it's sufficient to cast the literals for the first row only; SQL gets the hint from there. Oracle, on the other hand, needs all of the rows to be cast.

You now have a virtual table with a collection of test dates. You can now try out your age calculation:

```
--  PostgreSQL
    WITH dates(dob, today) AS (
        --  etc
    )
    SELECT
        dob, today,
        extract(year from age(today,dob)) AS age
    FROM dates;
--  MariaDB/MySQL
    WITH dates(dob, today) AS (
        --  etc
    )
    SELECT
        dob, today,
        timestampdiff(year,dob,current_timestamp) AS age
    FROM dates;

--  MSSQL
    WITH dates(dob, today) AS (
        --  etc
    )
    SELECT
        dob, today,
        datediff(year,dob,today) AS age
    FROM dates;
--  SQLite
    WITH dates(dob, today) AS (
        --  etc
    )
    SELECT
        dob, today,
        cast(
            strftime('%Y.%m%d', today) - strftime('%Y.%m%d', dob)
```

```
        as int) AS age
    FROM dates;
--  Oracle
    WITH dates(dob, today) AS (
        --  etc
    )
    SELECT
        dob, today,
        trunc(months_between(today,dob)/12) AS age
    FROM dates;
```

We've already noted in Chapter 4 how MSSQL gets the age wrong, and this is one way you can test this.

Using a Table Literal for Sorting

In Chapter 5, we noted the problem of sorting string values. Put simply, alphabetical order is rarely the best way to list items which are supposed to be sequential.

For example, if you have the day names of the week, the month names of the year, the colors of the rainbow, or even the names of the numbers ("One," "Two," "Three," etc.), sorting the strings in alphabetical order will just make things confusing.

In Chapter 5, we cheated by relying on a string position. Another, more resilient solution is to have a (virtual) table of values with their correct position.

In Chapter 8, we generated a summary of sales per weekday:

```
WITH
    data AS (...)
    summary AS (...)
SELECT
    weekday_number, total,
    100*total/sum(total) OVER()
FROM weekday_number
ORDER BY weekday_number;
```

The problem is that we've had to get the weekday number in order to sort this correctly. It would have been nicer to use the weekday name instead. We can then use an additional virtual table to sort the names.

First, let's redo the data CTE with the day name:

```
--  PostgreSQL, Oracle
    WITH data AS (
        SELECT to_char(ordered,'FMDay') AS weekday, total
        FROM sales
    )
--  MSSQL
    WITH data AS (
        SELECT datename(weekday,ordered) AS weekday, total
        FROM sales
    )
--  MariaDB/MySQL
    WITH data AS (
        SELECT date_format(ordered,'%W'), total
        FROM sales
    )
```

You'll notice that SQLite isn't included in the list. That's because it doesn't have a method of getting the weekday name. If you need it, you'll want the reverse technique in the next section.

The summary CTE will now group by the weekday name:

```
WITH
    data AS (
        SELECT
            ... AS weekday,
            total
        FROM sales
    ),
    summary AS (
        SELECT weekday, sum(total) AS total
        FROM data
        GROUP BY weekday
    )
--  etc
```

We'll now need a table literal with the days of the week as well as a sequence number.

sequence	weekday
1	Monday
2	Tuesday
3	Wednesday
4	Thursday
5	Friday
6	Saturday
7	Sunday

```
-- PostgreSQL, MariaDB (not MySQL), SQLite
   weekdays(sequence,weekday) AS (
       VALUES (1,'Monday'),(2,'Tuesday')          -- etc
   )
-- MySQL (not MariaDB)
   weekdays(sequence,weekday) AS (
       VALUES row(1,'Monday'), row(2,'Tuesday')   -- etc
   )
-- MSSQL
   weekdays(sequence,weekday) AS (
       SELECT * FROM (
           VALUES (1,'Monday'),(2,'Tuesday')      -- etc
       ) AS sq(a,b)
   )
-- Oracle
   weekdays(sequence,weekday) AS (
       SELECT 1,'Monday' FROM dual
       UNION ALL SELECT 2,'Tuesday' FROM dual     -- etc
   )
```

Finally, to do the sorting, you can join the summary CTE with the weekdays CTE and sort by the sequence number:

```
WITH
    data AS ( ),
    summary AS ( ),
    weekdays(dob, today) AS ( )
SELECT
    summary.weekday, summary.total,
    100*total/sum(summary.total) OVER()
FROM summary JOIN weekdays
    ON summary.weekday=weekdays.weekday
ORDER BY weekdays.sequence;
```

You should see something like this:

weekday	total	proportion
Monday	49304	15.081
Tuesday	45156.5	13.813
Wednesday	45959.5	14.058
Thursday	47528	14.538
Friday	42372.5	12.961
Saturday	48415.5	14.81
Sunday	48182.22	14.738

One advantage of this technique is that you can change the sequence numbering in the table literal, for example, to start on Wednesday if that suits you better.

By the way, if you're going to sort by weekday, or anything like it, very often, you might be better off saving the data in a permanent lookup table.

Using a Table Literal As a Lookup

We've already noted that SQLite has no function to get the day name of the week—only the day number. In any case, you may have other situations where the data you have isn't as friendly or as comprehensible as you would like.

One solution is to use a table literal to act as a lookup table.

For example, the vip table has a status level of 1, 2, or 3. You're supposed to realize that it means Gold, Silver, and Bronze. Here, we'll use a table literal to do just that.

First, we'll develop a CTE with the status names:

```
-- PostgreSQL, MariaDB (not MySQL), SQLite
   WITH statuses(status,name) AS (
       VALUES (1,'Gold'),(2,'Silver'),(3,'Bronze')
   )
-- MySQL (not MariaDB)
   WITH statuses(status,name) AS (
       VALUES row(1,'Gold'),row(2,'Silver'),row(3,'Bronze')
   )
-- MSSQL
   WITH statuses(status,name) AS (
       SELECT * FROM (
           VALUES (1,'Gold'),(2,'Silver'),(3,'Bronze')
       ) AS sq(a,b)
   )

-- Oracle
   WITH statuses(status,name) AS (
       SELECT 1,'Gold' FROM DUAL
       UNION ALL SELECT 2,'Silver' FROM DUAL
       UNION ALL SELECT 3,'Bronze' FROM DUAL
   )
```

We can now join the CTE to customers and vip tables:

```
WITH statuses(status,name) AS (
    -- etc
)
SELECT *
FROM
    customers
    LEFT JOIN vip ON customers.id=vip.id
    LEFT JOIN statuses ON vip.status=statuses.status
;
```

Again, the benefit is that you can change the status names on the fly.

You can also do the same sort of thing with author and customer genders. Another thing you can do with this technique is to translate from one set of names to another set of names.

You may be wondering why we don't include the full name of the gender or the vip status in the table itself. Remember that you should only record a piece of data once, and it should be the simplest version possible. Storing a value as a single character, as with the gender, or an integer, as with the vip status, reduces the possibility of data error or variation, and you can spell it out later when you want.

Splitting a String

If you have the courage to look in the script which generated the database, you'll find two recursive CTEs near the end:

```
--  Populate Genres
    INSERT INTO genres(genre)
    WITH split(bookid,genre,rest,genres) AS (
        ...
    )
    SELECT DISTINCT genre
    FROM split
    WHERE split.genre IS NOT NULL;
--  Populate Book Genres
    INSERT INTO bookgenres(bookid,genreid)
    WITH split(bookid,genre,rest,genres) AS (
        ...
    )
    SELECT split.bookid,genres.id
    FROM split JOIN genres ON split.genre=genres.genre
    WHERE split.genre IS NOT NULL;
```

The reason is purely pragmatic. There are thousands of book-genre combinations, and, instead of dumping the bookgenres table directly, it was more convenient to code the combined genres into a table and use the recursive CTEs to pull that apart. To leave the genres combined would have been wrong for all the reasons discussed in Chapter 3.

Here, we'll have a look at how this process works, by splitting a few sample strings.

We'll first take a simple string and put it in a table literal. For now, it will be a simple string with comma-separated values. Later, it will be more complex.

```
-- PostgreSQL, SQLite, MariaDB (not MySQL)
   WITH
       cte(fruit) AS (
           VALUES ('Apple,Banana,Cherry,Date,
               Elderberry,Fig')
       ),
-- MySQL (Not MariaDB)
   WITH
       cte(fruit) AS (
           VALUES row('Apple,Banana,Cherry,Date,
               Elderberry,Fig')
       ),
-- MSSQL
   WITH
       cte(fruit) AS (
           SELECT *
           FROM (VALUES ('Apple,Banana,Cherry,Date,
               Elderberry,Fig')) AS sq(a)
       ),
-- Oracle)
   WITH
       cte(fruit) AS (
           SELECT 'Apple,Banana,Cherry,Date,
               Elderberry,Fig' FROM dual
       ),
```

In order to make the code readable, the string has been split over two lines. *Don't do this in your real code!*

Some DBMSs don't like string literals with a line break inside. For those that will accept the line break, it will be part of the data, and we won't want that.

Be sure to write the string on one line, even if it's very long.

For the recursive CTE, we'll build two values: the individual item and a string containing the rest of the original string. The CTE can be called split:

```
WITH
    cte(fruit) AS (),
    split(fruit, rest) AS (

    )
```

The anchor member will get the first item from the string, up to the comma, and the rest, after the comma:

```
WITH
    cte(fruit) AS (),

--  PostgreSQL
    split(fruit, rest) AS (
        SELECT
            substring(fruit,0,position(',' in fruits)),
            substring(fruit,position(',' in fruits)+1)||','
        FROM cte
    )
--  MariaDB, MySQL
    split(fruit, rest) AS (
        SELECT
            substring(fruit,1,position(',' in fruits)-1),
            substring(fruit,position(',' in fruits)+1)||','
        FROM cte
    )
```

```
--  MSSQL
    split(fruit, rest) AS (
        SELECT
            cast(substring(fruit,0,charindex(',',fruits)) as varchar(255)),
            cast(substring(fruit,charindex(',',fruits)+1,255)+',' as varchar(255))
        FROM cte
    )
--  SQLite
    split(fruit, rest) AS (
        SELECT
            substring(fruit,0,instr(fruits,',')),
            substring(fruit,instr(fruits,',')+1)||','
        FROM cte
    )
--  Oracle
    split(fruit, rest) AS (
        SELECT
            substr(fruit,1,instr(fruits,',')-1),
            substr(fruit,instr(fruits,',')+1)||','
        FROM cte
    )
```

Note that for MSSQL we've had to cast the calculation to varchar(255) because of a peculiarity with string compatibility.

For the recursive member, we use the rest value. First, we get the string up to the first comma, which becomes the fruit value. Then, we get the rest of the string from the comma, which becomes the new value for rest:

```
WITH
    cte(fruit) AS (),
--  PostgreSQL
    split(fruit, rest) AS (
        SELECT ...
        UNION
        SELECT
            substring(rest,0,position(',' in rest)),
```

```
            substring(rest,position(',' in rest)+1)
        FROM cte WHERE rest<>''
    )
-- MariaDB, MySQL
    split(fruit, rest) AS (
        SELECT ...
        UNION
        SELECT
            substring(rest,1,position(',' in rest)-1),
            substring(rest,position(',' in rest)+1)
        FROM cte WHERE rest<>''
    )

-- MSSQL
    split(fruit, rest) AS (
        SELECT ...
        UNION ALL
        SELECT
            substring(rest,0,charindex(',', rest)),
            substring(rest,charindex(',', rest)+1,255)
        FROM cte WHERE rest<>''
    )
-- SQLite
    split(fruit, rest) AS (
        SELECT ...
        UNION
        SELECT
            substring(rest,0,instr(rest,',')),
            substring(rest,instr(rest,',')+1)
        FROM cte WHERE rest<>''
    )
-- Oracle
    split(fruit, rest) AS (
        SELECT ...
        UNION ALL
```

```
SELECT
    substr(rest,1,instr(rest,',')-1),
    substr(rest,instr(rest,',')+1)
FROM cte WHERE rest<>''
)
```

Note that we don't add a comma to the rest value this time: that was just to get started.

We have also added WHERE rest<>'' to the FROM clause. This is because we need to stop recursing when there's no more of the string to search.

You can now try it out:

```
WITH
    cte(fruit) AS (),
    split(fruit,rest) AS ()
SELECT * FROM split;
```

You should now see the following:

fruit	rest
Apple	Banana,Cherry,Date,Elderberry,Fig,
Banana	Cherry,Date,Elderberry,Fig,
Cherry	Date,Elderberry,Fig,
Date	Elderberry,Fig,
Elderberry	Fig,
Fig	[NULL]

Of course, we don't need to see the rest value in the output: it's just there so you can see its progress.

Splitting More Complex Data

So far, we've only split a simple string. We can do that with a more complex set of data. Here, we'll have a CTE with three rows and two columns:

name	list
colours	Red,Orange,Yellow,Green,Blue,Indigo,Violet
elements	Hydrogen,Helium,Lithium,Beryllium,Boron,Carbon
numbers	One,Two,Three,Four,Five,Six,Seven,Eight,Nine

The good news is that the process is nearly the same.

To begin with, we'll have a CTE with the table literal:

```
-- PostgreSQL, SQLite, MariaDB (not MySQL)
WITH
    cte(name,items) AS (
        VALUES
            ('colours','Red,Orange,...,Indigo,Violet'),
            ('elements','Hydrogen,Helium,...,Carbon'),
            ('numbers','One,Two,...,Eight,Nine')
    ),
-- MySQL (Not MariaDB)
WITH
    cte(name,items) AS (
        VALUES
            row('colours','Red,Orange,...,Indigo,Violet'),
            row('elements','Hydrogen,Helium,...,Carbon'),
            row('numbers','One,Two,...,Eight,Nine')
    ),
-- MSSQL
WITH
    cte(name,items) AS (
        SELECT *
        FROM (
            VALUES
                ('colours','Red,Orange,...,Indigo,Violet'),
                ('elements','Hydrogen,Helium,...,Carbon'),
                ('numbers','One,Two,...,Eight,Nine')
        ) AS sq(a,b)
    ),
```

```
-- Oracle)
   WITH
        cte(name,items) AS (
             SELECT 'colours','Red,Orange,...,Indigo,Violet'
                  FROM dual
             UNION ALL SELECT 'elements','Hydrogen,...,Carbon'
                  FROM dual
             UNION ALL SELECT 'numbers','One,Two,...,Eight,Nine'
                  FROM dual
        ),
```

(We've obviously abbreviated the lists to fit nicely on the page.)

For the anchor member, we start the same way as before, but we'll include the name of the list. It won't be involved in the split, but is useful for the output.

```
WITH
    cte(name, items) AS (),
-- PostgreSQL
    split(name, item, rest) AS (
        SELECT
             name,
             substring(items,0,position(',' in items)),
             substring(items,position(',' in items)+1)||','
        FROM cte
    )
-- MariaDB, MySQL
    split(name, list, rest) AS (
        SELECT
             name,
             substring(items,1,position(',' in items)-1),
             substring(items,position(',' in items)+1)||','
        FROM cte
    )
-- MSSQL
    split(name, list, rest) AS (
        SELECT
```

```
                name,
                cast(substring(items,0,charindex(',', items)) as varchar(255)),
                substring(items,charindex(',', items)+1,255)+','
            FROM cte
        )
--  SQLite
    split(name, list, rest) AS (
        SELECT
                name,
                substring(items,0,instr(items,',')),
                substring(items,instr(items,',')+1)||','
            FROM cte
        )
--  Oracle
    split(name, list, rest) AS (
        SELECT
                name,
                substr(items,1,instr(items,',')-1),
                substr(items,instr(items,',')+1)||','
            FROM cte
        )
```

As for the recursive member, again it's the same idea, with the name value included:

```
WITH
    cte(name, items) AS (),

--  PostgreSQL
    split(name, list, rest) AS (
        SELECT ...
        UNION
        SELECT
                name,
                substring(rest,0,position(',' in rest)),
                substring(rest,position(',' in rest)+1)
            FROM cte WHERE rest<>''
        )
```

```
--   MariaDB, MySQL
     split(name, list, rest) AS (
         SELECT ...
         UNION
         SELECT
             name,
             substring(rest,1,position(',' in rest)-1),
             substring(rest,position(',' in rest)+1)
         FROM cte WHERE rest<>''
     )
--   MSSQL
     split(name, list, rest) AS (
         SELECT ...
         UNION ALL
         SELECT
             name,
             cast(substring(rest,0,charindex(',', rest)) as varchar(255)),
             substring(rest,charindex(',', rest)+1,255)
         FROM cte WHERE rest<>''
     )
--   SQLite
     split(name, list, rest) AS (
         SELECT ...
         UNION
         SELECT
             name,
             substring(rest,0,instr(rest,',')),
             substring(rest,instr(rest,',')+1)
         FROM cte WHERE rest<>''
     )
--   Oracle
     split(name, list, rest) AS (
         SELECT ...
         UNION ALL
         SELECT
             name,
```

```
        substr(rest,1,instr(rest,',')-1),
        substr(rest,instr(rest,',')+1)
    FROM cte WHERE rest<>''
)
```

We can now put this to the test:

```
WITH
    cte(name, items) AS ()
    split(name, item, rest) AS ()
SELECT *
FROM split
ORDER BY name, item;
```

When it's all going, you should see something like the following:

name	item	rest
colours	Blue	Indigo,Violet,
colours	Green	Blue,Indigo,Violet,
colours	Indigo	Violet,
colours	Orange	Yellow,Green,Blue,Indigo,Violet,
colours	Red	Orange,Yellow,Green,Blue,Indigo,Violet,
colours	Violet	[NULL]
colours	Yellow	Green,Blue,Indigo,Violet,
elements	Beryllium	Boron,Carbon,
elements	Boron	Carbon,
elements	Carbon	[NULL]
elements	Helium	Lithium,Beryllium,Boron,Carbon,
elements	Hydrogen	Helium,Lithium,Beryllium,Boron,Carbon,
elements	Lithium	Beryllium,Boron,Carbon,
numbers	Eight	Nine,
numbers	Five	Six,Seven,Eight,Nine,

(continued)

name	item	rest
numbers	Four	Five,Six,Seven,Eight,Nine,
numbers	Nine	[NULL]
numbers	One	Two,Three,Four,Five,Six,Seven,Eight,Nine,
numbers	Seven	Eight,Nine,
numbers	Six	Seven,Eight,Nine,
numbers	Three	Four,Five,Six,Seven,Eight,Nine,
numbers	Two	Three,Four,Five,Six,Seven,Eight,Nine,

As you see the recursive CTE was able to work with multiple rows of data.

Summary

In this chapter, we had a closer look at using Common Table Expressions. A common table expression generates a virtual table that you can use later in the main query. In the past, you would make do with a subquery in the FROM clause.

The reason why you would use a CTE or a FROM subquery is that you might need to prepare data but you don't want to go to the trouble of saving it either in a view or a temporary table. CTEs are more ephemeral than temporary tables in that they are not saved at all.

CTEs have a number of advantages over FROM subqueries:

- You define the CTE *before* using it, making the query more readable and more manageable.

- You can chain multiple dependent or independent CTEs simply. If you wanted to do that with FROM subqueries, you would have to nest them, which gets unwieldy very quickly.

- CTEs can be recursive, so you can use them to iterate through data.

Simple CTEs

The simplest use of a CTE is to prepare data for further processing. Some uses include

- Defining a set of constant values, either as literals or as calculated values

- Preparing aggregate data, to be combined with non-aggregate queries

Parameter Names

A CTE is expected to have a name or alias for each column. You can define the names inside the CTE, or you can define them as part of the CTE definition.

Multiple CTEs

Some queries involve multiple steps. These steps can be implemented by chaining multiple CTEs.

Recursive CTEs

A recursive CTE is one which references itself. It can be used for iterating through a set of data.

Some uses of recursive CTEs include

- Generating a sequence of values

- Traversing a hierarchy through a self-join

- Splitting strings into smaller parts

Coming Up

So far, we've worked on a number of important major concepts. In the next chapter, we'll have a look at a few additional techniques you can use to work smarter with your database:

- **Triggers** allow you to automate a process whenever some of the data changes.

- **Pivot tables** are basically a two-dimensional aggregate query.

- **Variables** allow you to hold interim values when there's too much going on.

CHAPTER 10

More Techniques: Triggers, Pivot Tables, and Variables

Throughout the book, we've looked at pushing our knowledge and application of SQL a little further and explored a number of techniques, some new and some not so new.

When looking at some techniques, in particular, those involving aggregates and common table expressions, we also got a sense of pushing SQL deeper, with multitiered statements.

In this chapter, we'll go a little beyond simple SQL and explore a few techniques which supplement SQL. They're not directly related to each other, but they all allow you to do more in working with your data.

SQL triggers are small blocks of code which run automatically in some response to some database event. We'll look at how these work and how you would write one. In particular, we'll look at a trigger to automatically archive data which has been deleted.

Pivot tables are aggregates in two dimensions. They allow you to build summaries in both row and column data. We'll look at an example of preparing data to be summarized and how we produce a pivot table.

Variables are pieces of temporary data which can be used to maintain values between statements. They allow us to run a group of SQL statements, while they hold interim values which are passed from one statement to another. In this chapter, we'll look at using variables to hold temporary values while we add data to multiple tables.

© Mark Simon 2023
M. Simon, *Leveling Up with SQL*, https://doi.org/10.1007/978-1-4842-9685-1_10

Understanding Triggers

Sometimes, a simple SQL query isn't quite enough. Sometimes, what you really want is for a query to start off one or more additional queries. Sometimes, what you want is a **trigger**.

A trigger is a small block of code which will be run automatically when something happens to the database. There are various types of triggers, including

- DML (Data Manipulation Language) triggers run when some change is made to the data tables, as when the INSERT, UPDATE, or DELETE statements are executed.

- DDL (Data Definition Language) triggers run when changes are made to the structure of the database, such as when CREATE, ALTER, or DROP statements are executed.

- Logon triggers run when a user has logged in.

One reason you might use DDL or Logon triggers is if you want to track activity by storing this in a logging table.

Here, we're going to look more at a DML trigger.

Triggers can be used to fill in some shortcomings of standard DBMS behavior. Here are some examples which might call for a trigger:

- You might have an activity table which wants a date column updated every time you make a change. You can use a trigger to set the column for every insert or update.

- Suppose you have a rental table, where you enter a start and a finish date. You'd like the finish date to default to the start date if it isn't entered. SQL defaults aren't quite so clever, but you can set a trigger to set the finish date when you insert a new row.

- SQL has no auditing in the normal sense of the word. You can create a trigger to add some data to a logging table every time a row is added, updated, or deleted.

In this example, we're going to create a trigger to keep a copy of data which we're going to delete from the sales table.

In some of the preceding chapters, we've had to contend with the fact that in the sales table, some rows have NULLs for the ordered date/time. Presumably, those sales never checked out.

We've been pretty forgiving so far and filtered them out from time to time, but the time has come to deal with them. We can delete all of the NULL sales as follows:

```
-- Not Yet!
 DELETE FROM sales WHERE ordered IS NULL;
```

Note that there's a foreign key from the saleitems table to the sales table, which would normally disallow deleting the sales if there are any items attacked. However, if you check the script which generates the sample database, you'll notice the ON DELETE CASCADE clause, which will automatically delete the orphaned sale items.

When should you delete data? The short answer is never. The longer answer is more complicated. You would delete data that was entered in error, or you would delete test data when you've finished testing.

In this case, we're going to delete the sales with a NULL for the ordered date; we'll assume that the sale was never checked out and that the customer won't ever come back and finish it. However, we'll keep a copy of it anyway, just in case.

Most DBMSs handle triggers in a very similar way, but there are variations. We'll go over the basics first and then the details for individual DBMSs.

Some Trigger Basics

The basic syntax for creating a trigger is something like this:

```
CREATE TRIGGER something
ON some_table
BEFORE DELETE
BEGIN
    ...
END
```

None of the DBMSs do it exactly the same way, but it's roughly right:

- The trigger, of course, has a name: `CREATE TRIGGER something`.

- The trigger is attached to a table: `ON some_table`.

- The trigger is attached to an event.

The event is typically one of `BEFORE`, `AFTER`, or `INSTEAD OF`, followed by one of the DML statements. In this example, we want to do something with the old data before it's deleted.

For the sample trigger, we're going to copy the old data into a table called `deleted_sales`. This means that we're going to have to get to the data before it's vanished. The appropriate event is

```
BEFORE DELETE
```

It's going to be a little complicated, because we want to copy not only the data from the `sales` table but also from the `saleitems` table. We'll do that by concatenating those items into one string. You really shouldn't keep multiple items that way, but it's good enough for an archive, and you can always pull it apart if you ever need to.

The archive table looks something like this:

```
CREATE TABLE deleted_sales (
    id INT PRIMARY KEY,        -- Auto Incremented
    saleid INT,
    customerid INT,
    items VARCHAR(255),
    deleted_date TIMESTAMP  --  date/time
);
```

This table has already been created.

Preparing the Data to Be Archived

The trigger code will basically be an `INSERT` statement, inserting prepared values from the `sales` and `saleitems` tables. We'll prepare the values in a CTE:

```
-- PostgreSQL, MSSQL
   WITH cte AS (
       ...
   )
```

```
    INSERT INTO deleted_sales(saleid, customerid, items,
        deleted_date)
    SELECT saleid,customerid, items, current_timestamp
    FROM cte;

-- MariaDB/MySQL, SQLite, Oracle
    INSERT INTO deleted_sales(saleid, customerid, items, deleted_date)
    WITH cte AS (

        ...

    )
    SELECT saleid,customerid, items, current_timestamp
    FROM cte;
```

As you see, with some DBMSs you start with the CTE, as you would using a SELECT statement, while in others you start with the INSERT clause.

As for the CTE itself, we'll derive that from the data to be deleted.

For most DBMSs, each row to be deleted is represented in a virtual row called old (:old in Oracle). MSSQL instead has a virtual table called deleted.

If we were simply archiving from one table, we wouldn't need the CTE, and we could simply copy the rows with

```
-- Not MSSQL: FOR EACH ROW
    INSERT INTO deleted_sales
    VALUES(old.saleid, old.customerid, '...',
        current_timestamp);
-- MSSQL: deleted is a virtual table
    INSERT INTO deleted_sales
    SELECT saleid, customerid, '...', current_timestamp
    FROM deleted;
```

However, it's not so simple when there's another table involved. Here, the plan is to read the book ids and quantities from the other table and combine them using string_agg, group_concat, or listagg according to DBMS.

To generate the data, we'll use a join and aggregate the results:

```
WITH cte(saleid,customerid,items) AS (
    SELECT
        s.id, s.customerid,
        string_agg(si.bookid||':'||si.quantity,';')
    FROM sales AS s JOIN saleitems AS si ON s.id=si.saleid
    WHERE s.id=old.id
    GROUP BY s.id, s.customerid
)
```

The preceding sample is for PostgreSQL, but the others are nearly identical—just the variations in the `string_agg()` function, concatenation, and table aliases.

The `items` string will contain something like the following:

```
123:3;456:1;789:2
```

That is, one or more `bookid:quantity` items are joined with a semicolon.

If you do need to pull it apart, you can use the same techniques we used for splitting strings in Chapter 9. We can now go about creating the trigger.

Creating the Trigger

Now that we've covered the basics of how the trigger's going to work, we can write the actual code. For the most part, it will be pretty close to what you have seen earlier, with variations for each DBMS.

Once the trigger has been created, we can try it out with the following. The plan is to delete those sales without an `ordered` date:

```
-- Before
    SELECT * FROM sales order by id;
    SELECT * FROM saleitems order by id;
    SELECT * FROM deleted_sales order by id;
-- Delete with Trigger
    DELETE FROM sales WHERE ordered IS NULL;
```

```
-- After
   SELECT * FROM sales order by id;
   SELECT * FROM saleitems order by id;
   SELECT * FROM deleted_sales order by id;
```

We'll now go into the details for the individual DBMSs.

PostgreSQL Triggers

PostgreSQL has the least convenient form of trigger, in that you first need to prepare a **function** to contain the trigger code. A function is a named block of code, which can be called later at any time.

To prepare for the function and trigger, we can start with a few DROP statements:

```
DROP TRIGGER IF EXISTS archive_sales_trigger ON sales;
DROP FUNCTION IF EXISTS do_archive_sales;
```

The function will basically contain the code described earlier:

```
CREATE FUNCTION do_archive_sales() RETURNS TRIGGER
LANGUAGE plpgsql AS
$$BEGIN
    WITH cte(saleid,customerid,items) AS (
        SELECT
            s.id, s.customerid,
            string_agg(si.bookid||':'||si.quantity,';')
        FROM sales AS s JOIN saleitems AS si
            ON s.id=si.saleid
        WHERE s.id=old.id
        GROUP BY s.id, s.customerid
    )
    INSERT INTO deleted_sales(saleid, customerid,items,
        deleted_date)
    SELECT saleid, customerid, items, current_timestamp
    FROM cte;
    RETURN old;
END$$;
```

CHAPTER 10   MORE TECHNIQUES: TRIGGERS, PIVOT TABLES, AND VARIABLES

As you see, the function has the code for the CTE and for copying the data into the deleted_sales table. Here are a few points about the function itself:

- A function has a name (do_archive_sales) and returns a result of a certain type, in this case a TRIGGER.

- PostgreSQL has a number of alternative coding languages you can use to write a function, but the standard one is called plpgsql.

- Technically, a function definition is a string. However, using single quotes would interfere with single quotes inside the function definition. PostgreSQL allows an alternative string delimiter, in this case the $$ code. This is the most mysterious part of writing PostgreSQL functions.

Once you have the function in place, creating the trigger is simple:

```
CREATE TRIGGER archive_sales_trigger
    BEFORE DELETE ON sales
    FOR EACH ROW
    EXECUTE FUNCTION do_archive_sales();
```

You can now try it out.

MySQL/MariaDB Triggers

With MariaDB/MySQL, the trigger can be written in a single block. First, we'll write the code to drop the trigger:

```
DROP TRIGGER IF EXISTS archive_sales_trigger;
```

The basic form of the trigger code will be

```
CREATE TRIGGER archive_sales_trigger
    BEFORE DELETE ON sales
    FOR EACH ROW
BEGIN
    ...
END;
```

There's going to be a possible complication here. The trigger code includes a BEGIN ... END block which is to allow multiple statements on one block. At this point, MariaDB/MySQL isn't sure where the real end will be, so it's normal to change the end of the statement delimiter:

```
DELIMITER $$

CREATE TRIGGER archive_sales_trigger
    BEFORE DELETE ON sales
    FOR EACH ROW
BEGIN
    ...
END; $$

DELIMITER ;
```

Here, the delimiter is changed to $$. It doesn't have to be that, but it's a combination you're unlikely to use for anything else. The new delimiter is used to mark the end of the code and switched back to the semicolon after that.

After that, the trigger code is much as described:

```
DELIMITER $$
CREATE TRIGGER archive_sales_trigger
    BEFORE DELETE ON sales
    FOR EACH ROW
BEGIN
    INSERT INTO deleted_sales(saleid,customerid,items,deleted_date)
    WITH cte(saleid,customerid,items) AS (
        SELECT
            s.id, s.customerid,
            group_concat(si.bookid||':'||si.quantity SEPARATOR ';')
        FROM sales AS s JOIN saleitems AS si ON s.id=si.saleid
        WHERE s.id=old.id
        GROUP BY s.id, s.customerid
    )
```

```
    SELECT saleid,customerid,items,current_timestamp
    FROM cte;
END; $$

DELIMITER ;
```

You can now test your trigger.

MSSQL Triggers

MSSQL also has a simple, direct way of creating a trigger. However, there's a complicating factor, which we'll need to work around.

Before that, however, we'll add the code to drop the trigger:

```
DROP TRIGGER IF EXISTS archive_sales_trigger;
```

With other DBMSs, you create a BEFORE DELETE trigger to capture the data before it's gone. With MSSQL, you don't have that option: there's only AFTER DELETE and INSTEAD OF DELETE. In both cases, there is a virtual table called deleted which has the rows to be deleted.

The problem with AFTER DELETE is that, even though the deleted virtual table has the deleted rows from the sales table, it's too late to get the rows from the saleitems table, as they have also been deleted, but there's no virtual table for that.

For that, we'll take a different approach. We'll use an INSTEAD OF DELETE event, which is to say that MSSQL will run the trigger instead of actually deleting the data. The trick is to finish off the trigger by doing the delete at the end:

```
CREATE TRIGGER archive_sales_trigger
    ON sales
    INSTEAD OF DELETE AS
BEGIN
    ...
    DELETE FROM sales WHERE id IN(SELECT id FROM deleted);
END;
```

The deleted virtual table still has the rows which haven't actually been deleted, but were going to be before the trigger stepped in. All we need from that is the id to identify the sales which should be deleted at the end, together with the cascaded sale items.

The other complication is that MSSQL won't let you concatenate strings with numbers, so you'll have to cast the numbers as strings:

```
cast(si.bookid AS varchar)+':'+cast(si.quantity AS varchar)
```

In MSSQL, `varchar` is short for `varchar(30)`. It's much more than we need for the integers, but it will reduce to the actual size of the integer, and is easy to read.

The completed trigger code is

```
CREATE TRIGGER archive_sales_trigger
    ON sales
    INSTEAD OF DELETE AS
BEGIN
    WITH cte(saleid, customerid, items) AS (
        SELECT
            s.id, s.customerid,
            string_agg(cast(si.bookid AS varchar)+':'
                +cast(si.quantity AS varchar),';')
        FROM
            sales AS s
            JOIN saleitems AS si ON s.id=si.saleid
            JOIN deleted ON s.id=deleted.id
        GROUP BY s.id, s.customerid
    )
    INSERT INTO deleted_sales(saleid, customerid, items,
        deleted_date)
    SELECT saleid, customerid, items, current_timestamp
    FROM cte;
    DELETE FROM sales
    WHERE id IN(SELECT id FROM deleted);
END;
```

You can now delete your sales.

SQLite Triggers

Of all the DBMSs in this book, SQLite has by far the simplest and most direct version of coding a trigger.

First, we can write the code to drop the trigger:

```
DROP TRIGGER IF EXISTS archive_sales_trigger;
```

The code to create the trigger is almost identical to the discussion earlier:

```
CREATE TRIGGER archive_sales_trigger
    BEFORE DELETE ON sales
    FOR EACH ROW
BEGIN
    INSERT INTO deleted_sales(saleid, customerid, items,
        deleted_date)
    WITH cte(saleid, customerid, items) AS (
        SELECT
            s.id, s.customerid,
            group_concat(si.bookid||':'||si.quantity,';')
        FROM sales AS s JOIN saleitems AS si
            ON s.id=si.saleid
        WHERE s.id=old.id
        GROUP BY s.id, s.customerid
    )
    SELECT saleid, customerid, items,current_timestamp
    FROM cte;
END;
```

The FOR EACH ROW clause is optional, since in SQLite there's no alternative currently. However, it's included to make the point clear that the trigger applies to each row about to be deleted.

You can now test the trigger.

Oracle Triggers

Writing trigger code in Oracle is similar to the basic code outlined earlier, but there are a few complicating factors which we'll need to work around.

Before that, we can write the code to drop the trigger:

```
--  DROP TRIGGER archive_sales_trigger;
```

The code is commented out because Oracle doesn't support IF EXISTS.

The first complication is that the trigger code consists of multiple statements, so it's hard to tell which statements belong to a single block.

Oracle has an alternative statement delimiter which is used when you're trying to combine multiple statements:

```
/
CREATE TRIGGER archive_sales_trigger
    BEFORE DELETE ON sales
    FOR EACH ROW
BEGIN
    ...
END;
/
```

The forward slash (/) before and after the code defines the block. Everything between the slashes, including the statements terminated with a semicolon, will be treated as one block of code.

The second complication is that Oracle doesn't like making changes to the table doing the triggering. The solution is to tell Oracle that code is part of a separate transaction:

```
CREATE TRIGGER archive_sales_trigger
    BEFORE DELETE ON sales
    FOR EACH ROW
DECLARE
    PRAGMA AUTONOMOUS_TRANSACTION;
BEGIN
    ...
    COMMIT;
END;
```

A transaction is a group of changes which can be reversed ("rolled back") if something goes wrong. To keep the changes, however, you then need to use the COMMIT statement.

The rest of the code is much as discussed earlier:

```
/
CREATE TRIGGER archive_sales_trigger
    BEFORE DELETE ON sales
    FOR EACH ROW
DECLARE
    PRAGMA AUTONOMOUS_TRANSACTION;
BEGIN
    INSERT INTO deleted_sales(saleid, customerid, items, deleted_date)
    WITH cte(saleid,customerid,items) AS (
        SELECT
            s.id, s.customerid,
            listagg(si.bookid||':'||si.quantity,';')
        FROM sales s JOIN saleitems si ON s.id=si.saleid
        WHERE s.id=:old.id
        GROUP BY s.id, s.customerid
    )
    SELECT saleid, customerid, items, current_timestamp
    FROM cte;
    COMMIT;
END;
/
```

You can now test the trigger.

Pros and Cons of Triggers

The main role of triggers is to add behaviors to the database which aren't already there. For example, the DBMS already supports defaults and column constraints, so, though you could use triggers to do something similar, you should first check whether the built-in feature will do the job.

One example where triggers have come to the rescue is with earlier versions of Oracle. Other DBMSs have long implemented an autoincremented primary key, but Oracle didn't until more recent versions. In that case, you would use a trigger to maintain and use a separate sequence to fill in the primary key for the next inserted row.

We've already mentioned cases when you might want to provide a default value for a column which is more complex than the built-in default feature, or you might want to update a column automatically. Here, the trigger might be able to provide this extra functionality.

However, it's possible to get carried away with triggers. If there's a DML trigger on a table, then, every time you make any changes to the table data, there's always a little extra work, which might add an extra burden.

The other problem is that triggers might add a little more mystery to the database, especially to other users of the database. Every time you do something, something else happens. This can make troubleshooting a little trickier and make it a little harder to check that the data is correct.

Pivoting Data

One of the important principles of good database design is that each column does a different job. On top of that, each column is independent of the other columns. That's one reason why we put so much effort separating out the town details from the `customers` table in Chapter 2.

There are some situations, however, where this sort of design doesn't suit analysis. Take, for example, a typical ledger type of table:

date	description	food	travel	accommodation	misc
...	...	...	...	...	...
...	...	...	...	...	...

This is a layout that's very easy to understand and analyze. If you want to get the totals for a particular category, just add down the column. If you want to get the totals for a particular item, just add across. This sort of thing used to be done by hand until spreadsheets were invented to let the computer do all the hard work.

You may see this sort of design in database tables you come across. However, it's not a good design for SQL tables:

- Putting a value in one column precludes putting it in another: the columns are deeply dependent.

- You will end up with very many empty spaces.

- A new category means adding a new column to the table design. You may end up with a huge number of columns.

- The data is harder to analyze, because now you need calculate across *columns*: SQL aggregate functions are designed to aggregate across *rows*.

A better design would be

Date	description	category	amount
...	...	...	...
...	...	...	...

As a general rule, the categories should be in rows, not columns.

Still, it would be nice to be able to get the data from the second form into the first form. You can use it in presentations and possibly turn it into amazing charts.

Pivoting the Data

Generating the first type of result from the second is called pivoting the data. The idea is that the categories pivot (swing) around, from vertical to horizontal.

If you read the data in a spreadsheet program, you can pivot the data simply. However, you can also generate pivoted data directly within the database, though it's not quite so simple.

Generating the pivot table in the spreadsheet has the following advantages:

- It is more interactive, and you can easily change what is being pivoted and summarized.

- The spreadsheet will more automatically generate the categories; as you will see, this is not so convenient from within the database.

- You're only a step away from turning it into a chart.

On the other hand, using the database has the following advantages:

- The data is generated in a single environment.

- You can create a view to regenerate the pivot table at any time.

There are two main ways you can generate a pivot table in SQL:

- Manually: Using a GROUP BY clause, you can aggregate the data in multiple columns.

- MSSQL and Oracle have a built-in pivot table feature to do this for you. PostgreSQL also has one, but it's not built in and requires installation.

Since the purpose of pivoting data is to create summaries, you often need to use a grouped field or to group the values yourself. For example:

- You can use the existing state column which is a group of addresses.

- You can use date functions, such as month() to group dates by month.

- You can use string functions to extract a common part of a string.

The pivot table will look something like this:

row groups	column group	column group	column group
group 1	...	...	...
group 2	...	...	...
group 3	...	...	...

We're going to see how to pivot data from the sales and customers tables to get total sales by state and VIP categories. The result will look something like this:

State	gold	silver	bronze
...	...	...	...
...	...	...	...
...	...	...	...

In principle, you could transpose the table and have the VIP groups go down, with the states going across. This version, however, will look neater.

Manually Pivoting Data

As we've already seen before, you often need to prepare the data before you aggregate it. This particular summary will need data from four tables: `customers`, `towns`, `vip`, and `sales`. Fortunately, the `customerdetails` view already combines the `customers` and `towns` tables, so we can reduce the number to three.

All the preparation will be done in multiple CTEs:

```
WITH
    statuses AS (
        ...
    ),
    customerinfo AS (
        ...
    ),
    salesdata AS (

    )
...
```

- The `vip` table has a status number. The `statuses` CTE will be a table literal which allocates a name to the number.

- The `customerinfo` CTE will join the tables together and select the columns we want to summarize.

- The `salesdata` will be an aggregate query which will be a first step in our pivot table summary.

With those CTEs, we'll run another aggregate query which will result in our pivot table.

The status CTE is simple. We just need to match status numbers with names:

```
WITH
    statuses(status, statusname) As (
    --  PostgreSQL, SQLite, MariaDB (Not MySQL):
        VALUES (1,'Gold'), (2,'Silver'), (3,'Bronze')
    --  MySQL:
        VALUES row(1,'Gold'), row(2,'Silver'),
            row(3,'Bronze')
    --  MSSQL:
        SELECT * FROM (VALUES (1,'Gold'),(2,'Silver'),
            (3,'Bronze'))
    --  Oracle:
        SELECT 1,'Gold' FROM dual
        UNION ALL SELECT 2,'Silver' FROM dual
        UNION ALL SELECT 3,'Bronze' FROM dual
    )
```

The customerinfo CTE will join this to the customerdetails view and the vip table to get the id, state, and status name for the customers:

```
WITH
    statuses(status, statusname) AS (
        ...
    ),
    customerinfo(id, state, statusname) AS (
        SELECT customerdetails.id, state, statuses.statusname
        FROM
            customerdetails
            LEFT JOIN vip ON customerdetails.id=vip.id
            LEFT JOIN statuses ON vip.status=statuses.status
    )
SELECT *
FROM customerinfo;
```

If you test it now, you'll get something like this:

Id	state	statusname
407	NSW	Bronze
299	QLD	Gold
21	[NULL]	Gold
597	TAS	[NULL]
106	NSW	Gold
26	VIC	Gold
~ 303 rows ~		

At this point, you can group it by state or status name to see how many of each you have, but we're more interested in the total sales.

For that, we'll need to join the preceding with the `sales` table in another CTE:

```
WITH
    statuses(status, statusname) AS (
        ...
    ),
    customerinfo(id, state, statusname) AS (
        ...
    ),
    salesdata(state, statusname, total) AS (
        SELECT state, statusname, total
        FROM customerinfo JOIN sales
            ON customerinfo.id=sales.customerid
    )
SELECT *
FROM salesdata;
```

Again, testing what we have so far, we get

State	statusname	total
NSW	[NULL]	56
NSW	Silver	43.5
VIC	[NULL]	70
QLD	[NULL]	28
VIC	Gold	24.5
VIC	[NULL]	133
~ 5294 rows ~		

All of this is just to get the data ready. What we're going to do now is generate our group rows.

Obviously, you'll need an aggregate query, grouping by state. Normally, it would have looked something like this:

```
WITH
    statuses(status, statusname) AS (
        ...
    ),
    customerinfo(id, state, statusname) AS (
        ...
    ),
    salesdata(state, statusname, total) AS (
        ...
    )
SELECT state, sum(total)
FROM salesdata
GROUP BY state;
```

to give us this:

State	sum
WA	20274
ACT	6781.5
TAS	28193
VIC	79199.5
NSW	101889
NT	6151
QLD	53331.5
SA	30977.5

However, to get that ledger table appearance, we'll use aggregate filters to generate three separate totals:

```
WITH
    statuses(status, statusname) AS (
        ...
    ),
    customerinfo(id, state, statusname) AS (
        ...
    ),
    salesdata(state, statusname, total) AS (
        ...
    )
SELECT
    state,
    sum(CASE WHEN statusname='Gold' THEN total END) AS gold,
    sum(CASE WHEN statusname='Silver' THEN total END)
        AS silver,
    sum(CASE WHEN statusname='Bronze' THEN total END)
        AS bronze
FROM salesdata;
```

This finally gives you the following:

State	gold	silver	bronze
WA	213	1655	[NULL]
ACT	1272.5	[NULL]	[NULL]
TAS	4182.5	2203	2764.5
VIC	8190	5875	5752.5
NSW	11068.5	9319	10760.5
NT	[NULL]	[NULL]	339.5
QLD	5094	3522.5	10480
SA	644	1390.5	3362

You might be tempted to ask whether there's an easier way to do it. The answer is not really. The hard part was always going to be the preparation of the data for pivoting.

However, for a few DBMSs, the final step can be achieved with a built-in feature.

Using the Pivot Feature (MSSQL, Oracle)

MSSQL and Oracle both offer a non-standard pivot feature which simplifies generating the pivot table. It takes the following form:

```
SELECT ...
FROM ...
PIVOT (aggregate FOR column IN(columnnames)) AS alias
```

- The *aggregate* is the aggregate function you want to apply. In this case, it's sum(total).

- The *column* is the column whose values you want across the table. In this case, it's statusname.

- The *columnnames* is a list of values which will be the columns across the pivot table. In this case, it's Gold, Silver, Bronze.

- The alias is any alias you want to give. It's not used here, but it's required. The pivot table is, after all, a virtual table.

In our case, the pivot table will look like this:

```
WITH
    statuses(status, statusname) AS (
        ...
    ),
    customerinfo(id, state, statusname) AS (
        ...
    ),
    salesdata(state, statusname, total) AS (
        ...
    )
SELECT *
FROM salesdata
-- MSSQL:
PIVOT (sum(total) FOR statusname IN (Gold, Silver, Bronze))
    AS whatever
-- Oracle:
PIVOT (sum(total) FOR statusname IN ('Gold' AS Gold, 'Silver'
    AS Silver, 'Bronze' AS Bronze))
;
```

This is a little bit simpler than the filtered aggregates we used previously. However, note that there are some quirks with this technique.

The syntax for MSSQL and Oracle is not identical:

- In MSSQL, the column names list is a plain list of names. Also, note that the PIVOT clause requires an alias.

- In Oracle, the list of column names is a list of strings; however, they are aliased to prevent the single quotes from appearing in the names. The PIVOT clause itself does not require an alias.

You'll notice that the state doesn't make an appearance in the PIVOT clause; only the statusname and total. *Any column not mentioned in the* PIVOT *clause will appear as grouping rows.* You can have more complex pivot tables if there's more than one such column, but you need to make sure that the (virtual) table you want to pivot doesn't have any stray unwanted columns.

You'll also notice that the IN expression isn't a normal IN expression. To begin with, it's not a list of values, but a list of column names.

On top of that, you can't use a subquery to get the list of column names. You have to know ahead of time what the column names are going to be, and you'll have to type them in yourself.

Using the pivot feature is not quite as convenient as it might have been, but, if it's available, is still simpler than the filtered aggregates. However, you will still need to put in some effort in preparing your data first.

Using the Unpivot Feature

Both MSSQL and Oracle can reverse the process using UNPIVOT. Noting that a pivot table is denormalized, the UNPIVOT clause can give you a normalized result. The idea is that summaries which are spread across the table in category columns will appear down the table in rows.

The sample database doesn't include a table in pivot table form, and rightly so. However, we do have a ready-made pivot table in the preceding work. We can use that to illustrate the unpivot feature.

First, we'll need to wrap the final SELECT statement into another CTE:

```
WITH
    statuses(status, statusname) AS (
        ...
    ),
    customerinfo(id, state, statusname) AS (
        ...
    ),
    salesdata(state, statusname, total) AS (
        ...
    ), -- extra comma
    pivottable AS (
        SELECT *
        FROM salesdata
        PIVOT ...
    )
```

```
    SELECT *
    FROM pivottable
;
```

If you run this, you'll get the same result as before; we've just put the result into the pivottable CTE.

The next step is to add the UNPIVOT clause at the end of the SELECT statement:

```
WITH
    statuses(status, statusname) AS (
        ...
    ),
    customerinfo(id, state, statusname) AS (
        ...
    ),
    salesdata(state, statusname, total) AS (
        ...
    ),
    pivottable AS (
        ...
    )
    SELECT *
    FROM pivottable
    -- MSSQL:
    UNPIVOT (
        total FOR statuses IN (Gold,Silver,Bronze)
    ) AS w
    --  Oracle:
    UNPIVOT (
        total FOR statuses IN (Gold,Silver,Bronze)
    )
```

You should see something like

state	total	statuses
QLD	5532.5	Gold
QLD	3557.5	Silver
QLD	10937	Bronze
VIC	8352	Gold
VIC	6381.5	Silver
VIC	6023	Bronze
NSW	11526	Gold
NSW	9567	Silver
NSW	11941.5	Bronze
NT	349.5	Bronze
ACT	1387	Gold
TAS	4574	Gold
TAS	2459.5	Silver
TAS	2873.5	Bronze
SA	826.5	Gold
SA	1634.5	Silver
SA	3709.5	Bronze
WA	213	Gold
WA	1655	Silver

The UNPIVOT clause is even more mysterious than the PIVOT clause. The only column that's specifically mentioned is the statuses column, and, again, you need to list the possible values. From there, the DBMS magically works out that there is a state column, and whatever's left will appear in another column, which we have called total.

Working with SQL Variables

SQL is not a *programming* language. With a programming language, you code *how* to do a job in a series of steps. SQL is a *declarative* language in which you code what you want done, but leave it up to the DBMS to decide how to go about doing it.

Nevertheless, there are times when you need a job done in multiple steps, and having the ability to write your code in steps would come in handy. We saw this in Chapter 3, where adding a new sale involved multiple steps.

What was missing from the process in Chapter 3 was the ability to store interim values. That's what we're going to look at in this section.

Many DBMSs supply information about the current database environment in the form of special functions or system or global variables. Sometimes, these system variables can be set to new values using a SET command. That's *not* what we're looking at in this section. In this section, we're looking at variables that you create and set for your own use.

A **variable** is a temporary piece of data. Generally, you **declare** it before you use it and define its data type. You may set it then or, more typically, in a later step.

Typically, a variable is associated with a stored block of code called a **function** or a **procedure**, depending on the DBMS and what you're attempting to do in the code. In this section, we'll be working without storing the code.

The various DBMSs have slightly different processes for working with variables: ·

- PostgreSQL historically limited variables to stored code blocks (functions). However, in version 11 they introduced an **anonymous** (DO) block, which lets you write the code without storing the block. PostgreSQL variables must be declared with a data type.

- MariaDB/MySQL is most relaxed about variables, and you *don't* declare a variable before using it. Variable names are prefixed with the @ sign.

- MSSQL variables must be declared with a data type. They are prefixed with the @ sign.

- Oracle variables are declared with a data type.

SQLite is missing from this list, and that's because it doesn't support variables. SQLite is typically embedded in a host application. The assumption is that you're writing programming code for the host application. You can have all the additional variables and functionality you like there.

Code Blocks

If you're using a client which makes it easy to run one statement at a time, you may find it gets a little confused when working with blocks of multiple statements. It will be easier to work with if you surround your block with delimiters.

For the various DBMSs, the delimiters look like this:

```
--  PostgreSQL
    DO $$

       ...

    END $$;

--  MariaDB/MySQL
    DELIMITER $$

       ...

    $$
    DELIMITER ;

--  MSSQL
    GO

       ...

    GO

--  Oracle
    /

       ...

    /
```

In the end, you will probably just highlight all of the lines of code and run them together. That's what we recommend in trying the following code. *Don't try running just one line at a time.*

In the following code, we'll do what we did in Chapter 3 in adding a new sale. Then, we made a point of recording the new sale id, so that we could use it in subsequent statements. This time, however, we'll use variables to store interim values, so we can run the code in a single batch.

The code will broadly follow these steps:

1. Set up the data to be used.

2. Insert the sale.

3. Get the new sale id into a variable.

4. Insert the sale items, using the sale id.

5. Update the sale items with their prices, using the sale id.

6. Update the new sale with the total, using the sale id, of course.

While we're at it, we'll set a few other variables:

- A variable to store the customer's id

- A variable to store the ordered date/time

It would be nice to have another variable with the sale items. However, most DBMSs aren't adept at defining multivalued variables without a lot of extra fuss in defining custom data types to do the job. Here, we're trying to keep things simple.

What follows will be four similar versions of how to write the code block.

Updated Code to Add a Sale

The outline of the following code blocks will be basically the same:

- Define the variables

- Run the INSERT and UPDATE statements

- Test the results

We'll discuss the code for each of the main DBMSs.

Using Variables in PostgreSQL

Originally, PostgreSQL wouldn't let you do any of this outside a stored code block. From version 11 onward, you can use an **anonymous block**. If you're working with an older version of PostgreSQL, then you're out of luck.

The anonymous block is defined between DO ... END:

```
DO $$
    ...
END $$ ;
```

The $$ code is used to allow multiple statements to be treated as a single block. That way, the semicolon doesn't end up terminating the block prematurely.

Variables are declared inside a DECLARE section:

```
DO $$
DECLARE
    cid INT := 42;
    od TIMESTAMP := current_timestamp;
    sid INT;
END $$ ;
```

The variable names can be anything you like, but you run the risk of competing with column names in the following code. Some developers prefix the names with an underscore (such as _cid).

The sid variable is an integer which will be assigned later. The cid and od variables are for the customer id and ordered date/time. They are assigned from the beginning with the special operator :=.

The code proper is inside a BEGIN ... END block. It will be all of the code you used in Chapter 3, but run together. The important part is that the variable sid is used to manage the new sale id:

```
DO $$
DECLARE
    cid INT := 42;
    od TIMESTAMP := current_timestamp;
    sid INT;
BEGIN
```

397

```
INSERT INTO sales(customerid, ordered)
VALUES(cid, current_timestamp)
RETURNING id INTO sid;

INSERT INTO saleitems(saleid, bookid, quantity)
VALUES
    (sid,123,3),
    (sid,456,1),
    (sid,789,2);

UPDATE saleitems AS si
SET price=(SELECT price FROM books AS b
    WHERE b.id=si.bookid)
WHERE saleid=sid;

UPDATE sales
SET total=(SELECT sum(price*quantity)
    FROM saleitems WHERE saleid=sid)
WHERE id=sid;

END $$;
```

The sid variable gets its value from the RETURNING clause in the first INSERT statement. From there on, it's used in the remaining statements.

You can test the results using

```
SELECT * FROM sales ORDER BY id DESC;
SELECT * FROM saleitems ORDER BY id DESC;
```

You should see the new sale and sale items at the top.

Using Variables in MariaDB/MySQL

MariaDB/MySQL has the simplest approach to using variables. Outside of a stored function or procedure, you don't declare the variables or their types: you just go ahead and use them.

The other thing is that there's no strong concept of an anonymous code block, so defining one is really more of an organizational thing. We'll do that, even though it really makes no real difference:

```
DELIMITER $$
BEGIN

END; $$
DELIMITER ;
```

First, we'll assign a few variables:

```
DELIMITER $$
BEGIN
    SET @cid = 42;
    SET @od = current_timestamp;
    SET @sid = NULL;
END; $$
DELIMITER ;
```

Variables are prefixed with the @ character. This makes them a little more obvious and avoids possible conflict with column names.

The statement SET @sid = NULL; is unnecessary. Since you don't declare variables, we've included the statement just to make it clear that we'll be using the @sid variable a little later.

The whole code looks like this:

```
DELIMITER $$
BEGIN
    SET @cid = 42;
    SET @od = current_timestamp;
    SET @sid = NULL;    --  unnecessary; just to make clear

    INSERT INTO sales(customerid, ordered)
    VALUES(@cid, @od);

    SET @sid = last_insert_id();

    INSERT INTO saleitems(saleid,bookid,quantity)
    VALUES
        (@sid,123,3),
        (@sid,456,1),
        (@sid,789,2);
```

```
    UPDATE saleitems
    SET price=(SELECT price FROM books
        WHERE books.id=saleitems.bookid)
    WHERE saleid=@sid;

    UPDATE sales
    SET total=(SELECT sum(price*quantity)
        FROM saleitems WHERE saleid=@sid)
    WHERE id=@sid;
END;
$$

DELIMITER ;
```

Note the statement:

```
SET @sid = last_insert_id();
```

When you add a new row with an autogenerated primary key, you need to get the new value to use later. The `last_insert_id()` function fetches the most recent autogenerated value in the current session. You'll notice that it doesn't specify which table: that's why you need to call it immediately after the INSERT statement.

As you see, the rest of the code is generally the same as in Chapter 3, with the `@sid` variable used to manage the new sale id.

You can test the results using

```
SELECT * FROM sales ORDER BY id DESC;
SELECT * FROM saleitems ORDER BY id DESC;
```

You should see the new sale and sale items at the top.

Using Variables in MSSQL

MSSQL code can be written inside a block delimited with the GO keyword:

```
GO
    ...
GO
```

The GO keyword isn't actually a part of Microsoft's SQL language (or any other SQL, for that matter). It's actually an instruction to the client software to treat what's inside as a single batch and to run it as such. Some clients allow you to indent the keyword, and some allow you to add semicolons and comments on the same line, but the safest thing is not to indent it and not add anything else to the line.

Microsoft doesn't have a block to declare variables, but it does have a statement. To declare three variables, you can use three statements:

```
GO
    DECLARE @cid INT = 42;
    DECLARE @od datetime2 = current_timestamp;
    DECLARE @sid INT;
GO
```

or you can use a single statement with the variables separated by commas:

```
GO
    DECLARE
        @cid INT = 42,
        @od datetime2 = current_timestamp,
        @sid INT;
GO
```

Variables are prefixed with the @ character, which makes them easy to spot and easy to distinguish from column names.

The @sid variable is an integer which will be assigned later.

The rest of the code is similar to what we did in Chapter 3, but the new sale id will be managed in the @sid variable:

```
GO
    DECLARE @cid INT = 42;
    DECLARE @od datetime2 = current_timestamp;
    DECLARE @sid INT;

    INSERT INTO sales(customerid,ordered)
    VALUES(@cid, @od);

    SET @sid = scope_identity();
```

```
INSERT INTO saleitems(saleid,bookid,quantity)
VALUES
    (@sid,123,3),
    (@sid,456,1),
    (@sid,789,2);

UPDATE saleitems
SET price=(SELECT price FROM books
    WHERE books.id=saleitems.bookid)
WHERE saleid=@sid;

UPDATE sales
SET total=(SELECT sum(price*quantity)
    FROM saleitems WHERE saleid=@sid)
WHERE id=@sid;
GO
```

The @sid variable gets its value from the scope_identity() function. You'll notice that it doesn't specify which table: that's why you need to call it immediately after the INSERT statement. From there on, it's used in the remaining statements.

You can test the results using

```
SELECT * FROM sales ORDER BY id DESC;
SELECT * FROM saleitems ORDER BY id DESC;
```

You should see the new sale and sale items at the top.

Using Variables in Oracle

Oracle code blocks can be delimited with forward slashes:

```
/
```

```
/
```

When the time comes, the whole block will be run as a single batch.

Variables are declared inside a DECLARE section:

```
/
DECLARE
    cid INT := 42;
    od TIMESTAMP := current_timestamp;
    sid INT;
/
```

The variable names can be anything you like, but you run the risk of competing with column names in the following code. Some developers prefix the names with an underscore (such as _cid).

The sid variable is an integer which will be assigned later. The cid and od variables are for the customer id and ordered date/time. They are assigned from the beginning with the special operator :=.

The code proper is inside a BEGIN ... END block. It will be all of the code you used in Chapter 3, but run together. The important part is that the variable sid is used to manage the new sale id:

```
/
DECLARE
    cid INT := 42;
    od TIMESTAMP := current_timestamp;
    sid INT;
BEGIN
    INSERT INTO sales(customerid,ordered)
    VALUES(cid, od)
    RETURNING id INTO sid;

    INSERT INTO saleitems(saleid,bookid,quantity)
    VALUES (sid,123,3);
    INSERT INTO saleitems(saleid,bookid,quantity)
    VALUES (sid,456,1);
    INSERT INTO saleitems(saleid,bookid,quantity)
    VALUES (sid,789,2);

    UPDATE saleitems
    SET price=(SELECT price FROM books
```

```
        WHERE b.id=saleitems.bookid)
    WHERE saleid=sid;

    UPDATE sales
    SET total=(SELECT sum(price*quantity) FROM saleitems
        WHERE saleid=sid)
    WHERE id=sid;
END;
/
```

The sid variable gets its value from the RETURNING clause in the first INSERT statement. From there on, it's used in the remaining statements.

You can test the results using

```
SELECT * FROM sales ORDER BY id DESC;
SELECT * FROM saleitems ORDER BY id DESC;
```

You should see the new sale and sale items at the top.

Review

In this chapter, we've looked at a few additional techniques that can be used to get more out of our database.

Triggers

Triggers are code scripts which run in response to something happening in the database. Typically, these include INSERT, UPDATE, and DELETE events. Using a trigger, you can intercept the event and make your own additional changes to the affected table or another table. Some triggers can go further and work more closely with the DBMS or operating system.

We explored the concept by creating a trigger which responds to deleting from the sales table. In this case, we copied data from the sale and matching sale item into an archive table.

Different DBMSs vary in detail, but generally they follow the same principles:

- A trigger is defined for an event on a table.

- The trigger code has access to the data about to be affected.

- Using this data, the trigger code can go ahead and perform additional SQL operations.

Pivot Tables

A pivot table is a virtual table which summarizes data in both rows and columns. It's a sort of two-dimensional aggregate.

For the most part, raw table data isn't ready to be summarized this way. You would put some effort into preparing the data in the right form and making it available in one or more CTEs.

You can create a pivot table manually using a combination of two techniques:

- An aggregate query generates the vertical groups and the data to be summarized.

- A SELECT statement with aggregate filters generates a summary for each horizontal category.

MSSQL and Oracle both have a non-standard PIVOT clause which will, to some extent, automate the second process earlier. However, it still requires some input from the SQL developer to finish the job.

SQL Variables

In this chapter, we used variables to streamline the code, first introduced in Chapter 3, which adds a sale by inserting into multiple tables and updating them.

Most of the SQL we've worked with involved single statements. Some of those statements were effectively multipart statements with the use of CTEs to generate interim data.

In the case where you need more complex code to run in multiple statements, you may need to store interim values. These values are held in variables, which are temporary pieces of data.

In this chapter, we used variables for two purposes:

- To hold fixed values to be used in the code

- To store an interim value generated by some of the code

In most DBMSs, variables are declared and used within a block of code. In most cases, the variables and their values will vaporize after the code block is run. MariaDB/MySQL, however, will retain variables beyond the run.

SQLite doesn't support variables. It is expected that the hosting application will handle the temporary data that variables are supposed to manage.

Summary

Although you can go a long way with straightforward SQL statements and features, you can often get more out of your DBMS with some additional features:

- Triggers are used to run some code in response to some database event. They can be used to add some further processing to your database automatically.

- Pivot tables are virtual tables which provide a compact view of your summaries. You can generate a pivot table using a combination of aggregate queries, but some DBMSs offer a pivot feature to simplify the process.

- SQL variables are used to store temporary values between other SQL statements. They can be used to store interim values that can be used in subsequent statements.

Using what you've learned here and in previous chapters, you can build more complex queries to work with and analyze your database.

APPENDIX A

Cultural Notes

The sample database was based on the way we do things in Australia. This is pretty similar to the rest of the world, of course, but there are some details that might need clearing up.

Addresses and Phone Numbers

A standard address follows this pattern:

```
Street Number & Name
Town State Postcode
```

Australian addresses don't make much use of cities, which have a pretty broad definition in Australia.

Towns

Depending on how you define a town, there are about 15,000–20,000 towns in Australia.

In the sample database, town names have been deliberately selected as those occurring at least three times in Australia, though not necessarily in the sample.

States

Australia has eight geographical states. Technically, two of them are territories, since they don't have the same political features.

© Mark Simon 2023
M. Simon, *Leveling Up with SQL*, https://doi.org/10.1007/978-1-4842-9685-1

Each state has a two- or three-letter code.

Name	Code
Northern Territory	NT
New South Wales	NSW
Australian Capital Territory	ACT
Victoria	VIC
Queensland	QLD
South Australia	SA
Western Australia	WA
Tasmania	TAS

Postcodes

A postcode is a four-digit code typically, though not exclusively, associated with a town:

- Two adjacent towns may have the same postcode.

- A large town may have more than one postcode.

- A large organization may have its own postcode.

The postcode is closely associated with the state, though some towns close to the border may have a postcode from the neighboring state.

Phone Numbers

In Australia, a normal phone number has ten digits. For nonmobile numbers, the first two digits are an area code, starting with 0, which indicates one of four major regions. Mobile phones have a region code of 04.

There are also special types of phone numbers. Numbers beginning with 1800 are toll free, while numbers starting with 1300 are used for large businesses that are prepared to pay for them.

Shorter numbers starting with 13 are for very large organizations. Other shorter numbers are for special purposes, such as emergency numbers.

Australia maintains a group of fake phone numbers, and all of the phone numbers used in the database are, of course, fake. Don't waste your time trying to phone one.

Email Addresses

There are a number of special domains reserved for testing or teaching. These include `example.com` and `example.net`, which is why all of the email addresses use them.

This is true over the world.

Measurements and Prices and Currency

Australia uses the metric system, like most of the world. In particular, the sample database measures heights in centimeters. For those using legacy measurements, 1 inch = 2.54 cm.

For currency, Australia uses dollars and cents.

Prices on most things attract a **Goods and Services Tax** or GST to its friends. There are some exceptions to this, but not for anything in the sample database.

GST is a standard 10%.

In Australia, the GST is always expected to be displayed and is included in the asking price.

Dates

Short dates in Australia are in the day/month/year format, which can get particularly confusing when mixed with American and Canadian dates. It is for this reason that we recommend using the month name instead of the month number or, better still, the ISO8601 format.

APPENDIX B

DBMS Differences

This book covers writing code for the following popular DBMSs:

- PostgreSQL

- MySQL/MariaDB

- MSSQL: Microsoft SQL Server

- SQLite

- Oracle

Although there is an SQL standard, there will be variations in how well these DBMSs support them. For the most part, the SQL is 80–90% the same, with the most obvious differences discussed as follows.

As a rule, if there's a standard and non-standard way of doing the same thing, it's always better to follow the standard. That way, you can easily work with the other dialects. More importantly, you're future-proofing your code, as all vendors move toward implementing standards.

Writing SQL

In general, all DBMSs write the actual SQL in the same way. There are a few differences in syntax and in some of the data types.

© Mark Simon 2023
M. Simon, *Leveling Up with SQL*, https://doi.org/10.1007/978-1-4842-9685-1

Semicolons

MSSQL does not require the semicolon between statements. *However*, apart from being best practice to use it, Microsoft has stated that it will be required in a future version,[1] so you should always use one.

Data Types

All DBMSs have their own variations on data types, but they have a lot in common:

- SQLite doesn't enforce data types, but has general type affinities.

- PostgreSQL, MySQL/MariaDB, and SQLite support boolean types, while MSSQL and Oracle don't. MySQL/MariaDB tends to treat boolean values as integers.

Dates

- Oracle doesn't like ISO8601 date literals (yyyy-mm-dd). However, it is easy enough to get this to work. You can also use the to_date() function or the to_timestamp() function to accept different date formats.

- MariaDB/MySQL only accepts ISO8601 date literals. If you want to feed it a different format, you can use the str_to_date() function.

- SQLite doesn't actually have a date data type, so it's a bit more complicated. Generally, it's simplest to use a TEXT type to store ISO8601 strings, with appropriate functions to process it.

Case Sensitivity

Generally, the SQL language is case insensitive. However

- MySQL/MariaDB as well as Oracle may have issues with table *names*, depending on the underlying operating system.

---

[1] Microsoft's comment on semicolons: https://docs.microsoft.com/en-us/sql/t-sql/language-elements/transact-sql-syntax-conventions-transact-sql#transact-sql-syntax-conventions-transact-sql. TLDR: Semicolons are recommended and will be required in the future.

- Strings may well be case sensitive depending on the DBMS defaults and additional options when creating the database or table. By default

  - MSSQL and MySQL/MariaDB are case insensitive.

  - PostgreSQL, SQLite, and Oracle are case sensitive.

There's one more peculiarity in SQLite:

- Matching strings is case sensitive.

- Matching patterns (LIKE) is case insensitive.

Quote Marks

In standard SQL

- Single quotes are for 'values'.

- Double quotes are for "names".

However

- MySQL/MariaDB has two modes. In traditional mode, double quotes are also used for values, and you need the unofficial backtick for names. In ANSI mode, double quotes are for names.

- MSSQL also allows (and seems to prefer) square brackets for names. Personally, I discourage this, so it's not an issue.

Sorting (ORDER BY)

- Different DBMSs have different opinions on whether NULLs go at the beginning or the end.

- PostgreSQL, Oracle, and SQLite give you a choice.

Limiting Results

This is a feature omitted in the original SQL standards, so DBMSs have followed their own paths. However

- PostgreSQL, Oracle, and MSSQL all now use the OFFSET ... FETCH ... standard, with some minor variations.

- PostgreSQL, MySQL/MariaDB, and SQLite all support the non-standard LIMIT ... OFFSET ... clause. (That's right, PostgreSQL has both.)

- MSSQL also has its own non-standard TOP clause.

- Oracle also supports a non-standard row number.

Filtering (WHERE)

DBMSs also vary in how values are matched for filtering.

Unlike most DBMSs, SQLite will allow you to use an alias from the SELECT clause in the WHERE clause, which contradicts the standard clause order.

Case Sensitivity

This is discussed earlier.

String Comparisons

In standard SQL, trailing spaces are ignored for string comparisons, presumably to accommodate CHAR padding. More technically, shorter strings are right-padded to longer strings with spaces.

PostgreSQL, SQLite, and Oracle ignore this standard, so trailing spaces are significant. MSSQL and MySQL/MariaDB follow the standard.

Dates

Oracle's date handling is mentioned earlier. This will affect how you express a date comparison.

There is also the issue of how the ??/??/???? is interpreted. It may be the US d/m/y format, but it may not. It is *always* better to avoid this format.

Wildcard Matching

All DBMSs support the basic wildcard matches with the LIKE operator.

- PostgreSQL doesn't support wildcard matching with non-string data.

As for extensions to wildcards

- PostgreSQL, MySQL/MariaDB, and Oracle support regular expressions, but each one handles them differently.

- MSSQL doesn't support regular expressions, but does have a simple set of extensions to basic wildcards.

- SQLite has recently added native support for regular expressions (www.sqlite.org/releaselog/3_36_0.html).

Calculations

Basic calculations are the same, with the exceptions as follows. Functions, on the other hand, are very different.

Of the DBMSs listed earlier, SQLite has the fewest built-in functions, assuming that the work would be done mostly in the host application.

SELECT Without FROM

For testing purposes, all DBMSs except Oracle support SELECT without a FROM clause.

Oracle requires the dummy FROM dual clause. MariaDB/MySQL also allows you to use FROM dual, though it's rarely needed.

You can easily create your own DUAL table with the following code:

```
CREATE TABLE dual(
    dummy CHAR(1)
);
INSERT INTO dual VALUES('X');
```

Whether you would bother is another question.

415

Arithmetic

Arithmetic is mostly the same, but working with integers varies slightly:

- PostgreSQL, SQLite, and MSSQL will truncate integer division; Oracle and MySQL/MariaDB will return a decimal.

- Oracle doesn't support the remainder operator (%), but uses the mod() function.

Formatting Functions

Generally, they're all different. However

- PostgreSQL and Oracle both have the to_char() function.

- Microsoft has the format() function.

- SQLite only has a format() function, a.k.a. printf(), and is the most limited.

- MySQL/MariaDB has various specialized functions.

Date Functions

Again, all of the DBMSs have different sets of functions. However, for simple offsetting

- PostgreSQL and Oracle have the interval which makes adding to and subtracting from a data simple.

- MySQL/MariaDB has something similar, but less flexible.

- MSSQL relies on the dateadd() function.

- SQLite doesn't do dates, but it has some functions to process date-like strings.

Concatenation

This is a basic operation for strings:

- MSSQL uses the non-standard + operator to concatenate. Others use the || operator, with the partial exception of MySQL/MariaDB as follows.

- MySQL/MariaDB has two modes. In traditional mode, there is no concatenation operator; in ANSI mode, the standard || operator works.

- All DBMSs support the non-standard concat() function, with the exception of SQLite.

- Oracle treats the NULL string as an empty string. This is particularly noticeable when concatenating with a NULL which doesn't produce a NULL result as expected.

String Functions

Suffice to say that although there are some SQL standards

- Most DBMSs ignore them.

- Those that support them also have additional variations and functions.

This means that these examples will all require special attention.

Generally, the DBMSs support the popular string functions, such as lower() and upper() but sometimes in different ways. There is, however, a good deal of overlap between DBMSs.

Joining Tables

Everything is mostly the same. However

- Oracle doesn't permit the keyword AS for table aliases.

- SQLite doesn't support the RIGHT join.

Nobody knows why.

Aggregate Functions

The basic aggregate functions are generally the same between DBMSs. Some of the more esoteric functions are not so well supported by some.

PostgreSQL, Oracle, and MSSQL support an optional explicit GROUP BY () clause, which doesn't actually do anything important, but helps to illustrate a point. The others don't.

Manipulating Data

All DBMSs support the same basic operations. However

- Oracle doesn't support INSERT multiple values without a messy workaround, though there is talk of supporting it soon. MSSQL supports them, but only to a limit of 1000 rows, but there is also a less messy workaround for this limit. The rest are OK.

Manipulating Tables

All DBMSs support the same basic operations, but each one has its own variation on actual data type and autogenerated numbers.

Among other things, this means that the create table scripts are not cross-DBMS compatible.

- MSSQL has a quirk regarding unique indexes on nullable columns, for which there is a workaround.

Autoincremented Primary Keys

Inserting your own value into an autoincremented primary key may require you to make adjustments once you've finished. Typically, this will cause the DBMS to start autoincrementing from the right value:

- For PostgreSQL, you reset the underlying sequence after inserting the data. For example:

```
SELECT setval(pg_get_serial_sequence('customers',
    'id'), max(id))
FROM customers;
```

- For Oracle, alter the table you've just added data to. For example:

```
ALTER TABLE customers
MODIFY ID GENERATED BY DEFAULT AS IDENTITY
START WITH LIMIT VALUE;
```

- For MSSQL, switch between allowing your own values and autoincremented values. For example:

```
SET IDENTITY_INSERT customers ON;
--     INSERT statements ...
SET IDENTITY_INSERT customers OFF;
```

The other DBMSs seem to cope.

Other Quirks and Variations

Here is a miscellaneous collection of differences, some interesting and some fairly important.

PostgreSQL Quirks and Variations

PostgreSQL allows you to cast the strings yesterday, today, and tomorrow as dates.

Microsoft Quirks and Variations

Microsoft has this thing about CREATE statements, such as CREATE VIEW, being the only one in a batch. You define a batch with the GO keyword:

```
GO
    CREATE something AS
        ...
    ;
GO
```

That doesn't include CREATE TABLE, which will happily mix in with the rest of the statements.

Oracle Quirks and Variations

You can't mix a plain star (*) with column names in a SELECT clause. You need to qualify the star:

```
SELECT
    id, customers.*
FROM customers;
```

Casting a timestamp, which is a combined date/time, to a date doesn't cast it to just a date: it still retains the time component.

Instead, you should use trunc(). This will still have a time component, but it's set to midnight.

MariaDB/MySQL Quirks and Variations

If you want to mix the star (*) with column names in a SELECT clause, you need to put the star first:

```
SELECT *, id        -- NOT id, *
FROM customers;
```

For the OFFSET ... LIMIT ... clause, which fetches a limited number of rows, the OFFSET value cannot be calculated.

As you know, in a GROUP BY query, you can only select aggregates or what's in the GROUP BY clause. With MariaDB/MySQL, that won't work if the GROUP BY column is calculated. You really should be using CTEs anyway.

Don't forget to set your session to ANSI mode to have MariaDB/MySQL behave like the rest in the use of double quotes and concatenation:

```
SET session sql_mode = 'ANSI';
```

APPENDIX C

Using SQL with Python

Python has become a popular programming language in both scientific and data analysis spheres. In this appendix, we're going to look at how to connect your Python program to an existing database and both read from and write to the database.

If you're reading this, we'll assume that you're familiar with programming in Python, though not necessarily an expert.

In particular, we'll assume that, apart from the basics, you know about collections such as tuples, lists, and dictionaries. Of course, you'll be familiar with creating a function. You'll also need to know about installing and importing modules.

Before any of this can happen, however, you will probably have to install the appropriate module.

Once you've done that, we'll go through the following steps:

1. `import` the database module.

2. Make a connection to the database and store the **connection** object and a corresponding **cursor** object.

3. Run your SQL and process the results.

4. Close the connection.

A connection object represents a connection to the database, and you can use it to manage your database session.

More importantly, a cursor object is what you'll use to send SQL to the database and to send and receive the data involved. The connection object also has some data manipulation methods, but what they really do is create a cursor and pass on the rest of the work to a cursor.

© Mark Simon 2023
M. Simon, *Leveling Up with SQL*, https://doi.org/10.1007/978-1-4842-9685-1

Installing the Database Connector Module

For most DBMSs, installing the appropriate module is easy enough, once you work out the name of the module. To install the module, you'll need to use the `pip` program; sometimes, it's called `pip3` to reflect the current Python version.

The exception is with MSSQL, which might require a bit more work, especially if you're doing this on Macintosh or Linux. We'll look at that after the others.

The other exception is for SQLite. The module, called `sqlite3`, is already packaged with Python, so there's one less thing you need to do.

For the others, in your shell or command line, enter one of the following:

```
#   MariaDB/MySQL
    pip3 install mysql-connector-python
#   PostgreSQL
    pip3 install psycopg2-binary
#   Oracle
    pip3 install oracledb
```

The module for the preceding MariaDB and MySQL is the same. However, there is a dedicated MariaDB module if you need more specialized features.

Installing the MSSQL Module on Windows

For MSSQL, installing on Windows isn't too hard:

```
#   MSSQL (Windows)
    pip3 install pyodbc
```

The `pyodbc` module requires ODBC (Open Database Connectivity) drivers to do the job. On Windows, this will already be installed, especially if you've also installed SQL Server. However, you will need to get the name of the driver.

In Python, run the following:

```
import pyodbc
print(pyodbc.drivers())
```

You'll see a collection of one or more drivers. The one you want will be something like

```
ODBC Driver 18 for SQL Server
```

depending on the version.

Installing the MSSQL Module on Macintosh or Linux

On Macintosh or Linux, you'll need to install the ODBC (Open Database Connectivity) drivers yourself. You can get the instructions at https://learn.microsoft.com/en-us/sql/connect/python/pyodbc/step-1-configure-development-environment-for-pyodbc-python-development?view=sql-server-ver16#macos.

Here are the extra steps for Macintosh. You'll first need to install **Homebrew** (https://brew.sh/), which is a package manager which enables you to install all sorts of terminal applications:

```
/bin/bash -c "$(curl -fsSL https://raw.githubusercontent.com/Homebrew/
    install/master/install.sh)"
```

The command is too long to fit on this page. You should enter the command on one line, with no break or spaces in the URL.

Once you've got Homebrew installed, you can use it to install the correct driver for MSSQL:

```
brew tap microsoft/mssql-release \
    https://github.com/Microsoft/homebrew-mssql-release
```

Again, the command is too long to fit. You can write it on two lines as long as the first line ends with a backslash; otherwise, write it on one line without the backslash.

But wait, there's more. You then need to install the next part, at the same time accepting the license agreement:

```
HOMEBREW_ACCEPT_EULA=Y brew install msodbcsql18 mssql-tools18
```

Next, it's recommended that you install another driver:

```
brew install unixodbc
```

Now, you can install the module. You may have trouble installing it simply, especially if you're using an M1 Macintosh, so it's safer to run this:

```
pip3 install --no-binary :all: pyodbc
```

After this, you will need to get the name of the driver.

In Python, run the following:

```
import pyodbc
print(pyodbc.drivers())
```

You'll see a collection of one or more drivers. The one you want will be something like

```
ODBC Driver 18 for SQL Server
```

depending on the version.

Creating a Connection

Overall, to make a connection and cursor to the database, your code will look something like this:

```
import dbmodule
connection = dbmodule.connect(...)
cursor = connection.cursor()

connection.close()
```

where dbmodule is the relevant module for the DBMS. Specifically, for the various DBMSs, the code will be as follows.

Connecting to SQLite

The relevant module for SQLite is called sqlite3. After importing the module, you need to make the connection to the database.

SQLite databases are in simple files. You'll find there are no further credentials to worry about, since that's supposed to be handled in the host application. All you need to do is to reference the file.

To connect to SQLite

```
import sqlite3
connection = sqlite3.connect(file)  #   path name of the file
cursor = connection.cursor()
```

The file string is the full or relative path name of the SQLite file.

Connecting to MSSQL

The module or MSSQL is called pyodbc. In principle, it can be used for any database which supports ODBC.

A connection in MSSQL can be a string with all of the connection details. This string is called a DSN—a Data Source Name. However, for readability and maintainability, it's easier to add the details as separate function parameters. In general, it looks like this:

```
import pyodbc
connection = pyodbc.connect(
    driver='ODBC Driver 18 for SQL Server',
    TrustServerCertificate='yes',
    server='...',
    database='bookshop',
    uid='...',
    pwd='...'
)
cursor = connection.cursor()
```

As for the parameters

- driver is the name of the current database driver you would have obtained earlier. At the time of writing, the latest version is ODBC Driver 18 for SQL Server.

- TrustServerCertificate is to allow a connection to another server. This more or less tells the Python application to trust the server.

- server is the name or IP address of your database server. The standard port number is 1433. If you need to change the port number, you can add it to the server address:

  ```
  server='...,1432
  ```

- database is the name of the database.

- uid is the username, while pwd is the password.

Connecting to MariaDB/MySQL

The relevant module to connect to MariaDB/MySQL is called mysql.connector. To connect to the database, you will need to indicate which server and database, as well as your username and password:

```
import mysql.connector
connection = mysql.connector.connect(
    user='...',
    password='...',
    host='...',
    database='bookshop'
)
cursor = connection.cursor()
```

The host is typically the IP address of the database server. The standard port number is 3306. If you need to change the port number, you can add it as another parameter: port=3305.

Connecting to PostgreSQL

The module to connect to PostgreSQL is called psycopg2. To connect to the database, you will need to indicate which server and database, as well as your username and password:

```
import psycopg2
connection = psycopg2.connect(
    database='...',
    user='...',
    password='...',
    host='...'
)
cursor = connection.cursor()
```

The host is typically the IP address of the database server. The standard port number is 5432. If you need to change the port number, you can add it as another parameter: port=5433.

Connecting to Oracle

The module to connect to Oracle is called oracledb. To connect to the database, you will need to indicate which server and database, as well as your username and password:

```
import oracledb
connection = oracledb.connect(
    user='...',
    password='...',
    host='...',
    service_name='...'
)
cursor = connection.cursor()
```

The host is typically the IP address of the database server. The standard port number is 1521. If you need to change the port number, you can add it as another parameter: port=1522.

Fetching from the Database

Having made the connection, the next step is to send some SQL to the database and process its results.

The SQL statement is set in a simple string:

```
sql = 'SELECT * FROM customers'
```

You then use the connection object to execute the statement:

```
connection.execute(sql)
```

Before we process the data, we'll want to get a list of column names. This information is available in the cursor.description object. The cursor.description object is a tuple of tuples, one for each column. The data inside each of the tuples may include information about the type of data, but that's not available for all DBMS connections.

The column names will be the first item of each tuple. We can gather the names using a list comprehension:

```
columns = [i[0] for i in cursor.description]
```

This adds the first member of each tuple to the `columns` list.

The data from the `SELECT` statement will be available from the `cursor` object. The object includes methods to fetch one or more rows, but can also be iterated to fetch the rows.

You can iterate through the cursor as follows:

```
for row in cursor:
    print(row)
```

Each row will be a tuple of values. You'll recall that a tuple is a simple immutable collection of values, so, among other things, the values don't have a name.

You can combine the column names with each tuple using Python's `zip` function, which has nothing to do with zipping a file.

The `zip` function will take two collections and return a collection of tuples, each with an element from the first collection and an element from the second collection:

```
zip(columns,row)
```

Here, the result will be a collection of tuples with the first member being a column name and the second member being a corresponding value from the row. Technically, it's not a collection, but an iterator which is close enough for the next step.

Our next step will be to turn that into a dictionary object, using the first member of each tuple as keys for the second member of the tuple.

This will produce a set of dictionary objects:

```
data = []
for row in cursor:
    data.append(dict(zip(colums,row)))
print(data)
```

You can, of course, decide what to do with the data yourself.

When you've finished, you should close the connection:

```
connection.close()
```

The whole process looks like this:

```
import ...
connection = ... . connect(...)

sql = 'SELECT * FROM customers'
connection.execute(sql)

columns = [i[0] for i in cursor.description]

data = []
for row in cursor:
    data.append(dict(zip(colums,row)))
print(data)

connection.close()
```

Using Parameters in the Query

Once we've tested with a simple SQL query, we can try something a little more interesting. Let's look for one particular customer, say, customer 42. We can try changing the sql string to the following:

```
sql = 'SELECT * FROM customers WHERE id=42'
```

That will work, but it's too hard-coded to be useful. Instead, we'll get the customer id from the user:

```
customerid = input('Customer Number: ')
```

To put the customer id into the sql string, we could try something like this:

```
#   This is a bad idea:
customerid = input('Customer Number: ')
sql = f'SELECT * FROM customers WHERE id={customerid}'
#sql = 'SELECT * FROM customers WHERE id={0}'
        .format(customerid)
cursor.execute(sql)
```

Modern Python supports the so-called f-string earlier. Alternatively, you could use the more traditional `format()` string method.

The problem is that now you've opened up the query to user input. If, instead of entering the number 42, the user had entered

```
42 OR 1=1
```

the resulting string would be

```
SELECT * FROM customers WHERE id=42 OR 1=1
```

and they would have got the lot.

It could get worse. We don't have passwords here, but if we did, you can see how a user might override the password check with some carefully crafted input.[2]

Squeezing additional SQL code into the original code is called **SQL Injection**, and you run the risk of compromising and even losing your data. It works because the DBMS doesn't get the SQL string until after the extra code has been added, so it has no idea what's genuine and what's not. When it comes to interpreting the string, it's too late.

To safely handle including user input, you need to interpret the string *before* it gets the data. This is normally referred to as a **prepared** statement. To do this in Python, you need two steps:

- Create the SQL string with placeholders instead of data.

- Execute the SQL string with the data afterward.

Different DBMSs have different placeholders. Here is how you can create your SQL strings:

```
#   SQLite, MSSQL (use ?)
sql = 'SELECT * FROM customers WHERE id=?'
#   PostgreSQL, MariaDB/MySQL (use %s)
sql = 'SELECT * FROM customers WHERE id=%s'
#   Oracle (:named or :numbered)
sql = 'SELECT * FROM customers WHERE id=:0'
```

---

[2] Don't even think about storing passwords simply in a database table. This isn't the place to discuss how to manage user data safely, but storing plain passwords is very dangerous and irresponsible.

- SQLite and MSSQL use **?** for placeholders.

- PostgreSQL and MariaDB/MySQL use **%d** for placeholders.

- Oracle uses the colon followed by a name or a number.

Some DBMSs also allow variations on the preceding steps, such as using placeholder names. However, these simple placeholders will do well enough.

You can then add your data in the form of a tuple:

```
(customerid,)
```

Remember that a tuple with a single value requires a comma at the end.

The code should now look like

```
$sql = '...'  #   SELECT with placeholders
customerid = input('Customer Number: ')
cursor.execute(sql, (customerid,))
```

You can see that the tuple with values is added as a second parameter to the execute() method.

Adding a New Sale

Of course, you can also use Python to do something more complex. This time, we're going to add a new sale, as we have already done before.

Remember the steps:

1. Add a new sale.

2. Get the new sale id.

3. Add the books.

4. Get the book prices.

5. Put the total into the sale.

For simplicity, we can create separate SQL strings for the main steps:

```
insertsale = '...'    #   Add new sale
insertitems = '...'   #   Add sale items with books
```

```
updateitems = '...'    #    Update sale items with book prices
updatesale = '...'     #    Update sale with total
```

We'll get to those strings in a moment. Before we do, we need to look out for the new sale id.

In SQL, there are two main methods of getting a newly generated id:

- Return it from the INSERT statement.

- Fetch it in a separate step.

The first method is better, but isn't supported by all DBMSs at this stage. We'll need to take that into account with the first SQL string.

The other thing is that we'll include placeholders in these strings. That's not strictly necessary at this point, since we're not including user input. However, it's safer and makes adding the values easier.

To make the code a little more reusable, we'll wrap it inside a function:

```
def addsale(customerid, items, date):
    insertsale = '...'     #    Add new sale
    insertitems = '...'    #    Add sale items with books
    updateitems = '...'    #    Update sale items with book prices
    updatesale = '...'     #    Update sale with total

    return saleid
```

The customerid will be a simple integer. The items will be a list of dictionaries, which we'll describe later. The date will be a date object.

We don't really need to return the saleid, but it doesn't hurt, and it might come in handy later.

The SQL Strings

The actual strings all vary subtly between DBMSs, so we'll treat them per DBMS. The main considerations are how the new id is returned and how placeholders are represented.

Some of the strings are long; we've used multiline strings for readability. In Python, multiline strings have triple quote characters:

```
multiline = '''
Multi
Line
String
'''
```

The other thing is whether you use single or double quotes. Many developers use double quotes both for single-line strings and multiline strings. In this appendix, we're using single quotes. It doesn't matter, as long as you're consistent.

SQL Strings for PostgreSQL

PostgreSQL can return the new id from a RETURNING clause in the INSERT statement. Later, we'll fetch that value.

The strings look like this:

```
insertsale = '''
    INSERT INTO sales(customerid, ordered) VALUES(%s,%s) RETURNING id;
'''

insertitems = '''
    INSERT INTO saleitems(saleid, bookid, quantity) VALUES(%s,%s,%s);
'''

updateitems = '''
    UPDATE saleitems SET price=(SELECT price FROM books WHERE
        books.id=saleitems.bookid) WHERE saleid=%s;
'''

updatesale = '''
    UPDATE sales SET total=(SELECT sum(price*quantity)
        FROM saleitems WHERE saleid=%s) WHERE id=%s;
'''
```

These are mostly the statements we used earlier in the book.

SQL Strings for SQLite

SQLite doesn't return the new idea from the INSERT statement, so we'll have to get that later using a different technique. The strings look like this:

```
insertsale = '''
    INSERT INTO sales(customerid, ordered) VALUES(?,?)
    RETURNING id;
'''

insertitems = '''
    INSERT INTO saleitems(saleid, bookid, quantity)
    VALUES(?,?,?);
'''

updateitems = '''
    UPDATE saleitems SET price=(SELECT price FROM books
        WHERE books.id=saleitems.bookid) WHERE saleid=?;
'''

updatesale = '''
    UPDATE sales SET total=(SELECT sum(price*quantity)
        FROM saleitems WHERE saleid=?) WHERE id=?;
'''
```

These are mostly the statements we used earlier in the book.

SQL Strings for MSSQL

MSSQL can also return the id from the INSERT statement, but it uses a non-standard OUTPUT clause. It's awkward to use in a simple INSERT statement, but works well when used in Python code:

```
insertsale = '''
    INSERT INTO sales(customerid, ordered)
    OUTPUT inserted.id VALUES(?,?);
'''

insertitems = '''
    INSERT INTO saleitems(saleid, bookid, quantity)
    VALUES(?,?,?);
'''
```

```
updateitems = '''
    UPDATE saleitems SET price=(SELECT price FROM books
        WHERE books.id=saleitems.bookid) WHERE saleid=?;
'''

updatesale = '''
    UPDATE sales SET total=(SELECT sum(price*quantity)
        FROM saleitems WHERE saleid=?) WHERE id=?;
'''
```

Apart from the OUTPUT clause, these are basically the statements we used earlier.

SQL Strings for MariaDB/MySQL

MariaDB/MySQL doesn't return the new id from the INSERT statement, so we'll have to get that later using a different technique. The strings look like this:

```
insertsale = '''
    INSERT INTO sales(customerid, ordered) VALUES(%s,%s);
'''

insertitems = '''
    INSERT INTO saleitems(saleid, bookid, quantity)
    VALUES(%s,%s,%s);
'''

updateitems = '''
    UPDATE saleitems SET price=(SELECT price FROM books
        WHERE books.id=saleitems.bookid) WHERE saleid=%s;
'''

updatesale = '''
    UPDATE sales SET total=(SELECT sum(price*quantity)
        FROM saleitems WHERE saleid=%s) WHERE id=%s;
'''
```

These are mostly the statements we used earlier in the book.

SQL Strings for Oracle

Oracle can return the new id from the INSERT statement, though it uses an unusual syntax: you'll need to include a placeholder for the returned value.

There is also a complication with the date/time. Oracle is very fussy about date/time format and will probably reject the format used by Python. You'll need to include a to_timestamp() function which will convert the input to something Oracle can handle. It looks like this:

```
to_timestamp( ... , 'YYYY-MM-DD HH24:MI:SS')
```

Oracle can use named or numbered placeholders. We'll use numbered placeholders because it's simple:

```
insertsale = '''
    INSERT INTO sales(customerid, ordered)
    VALUES(:1, to_timestamp(:2,'YYYY-MM-DD HH24:MI:SS'))
    RETURNING id INTO :3
'''
insertitems = '''
    INSERT INTO saleitems(saleid, bookid, quantity)
    VALUES(:1,:2,:3)
'''
updateitems = '''
    UPDATE saleitems SET price=(SELECT price FROM books
        WHERE books.id=saleitems.bookid) WHERE saleid=:1
'''
updatesale = '''
    UPDATE sales SET total=(SELECT sum(price*quantity)
        FROM saleitems WHERE saleid=:1) WHERE id=:2
'''
```

Watch out for this quirk: *you cannot end the statements with a semicolon*! If you do, you'll get an error message: SQL command not properly ended, which is somewhat counterintuitive.

Note that the insertsale string includes an expression with single quotes. That's OK if the string is delimited with triple characters. If you're writing it on one line, you might need to use double quotes for the string.

Adding the Sale

Once we have created the strings, the next step will be to execute the first SQL statement to add the sale and then to fetch the resulting sale id.

Again, this varies by DBMS, but the technique is simple. The first step is to execute the `insertsale` query, with a tuple of the customer id and the date, which come from the `addsale()` function parameters:

```
#   Not Oracle
cursor.execute(insertsale, (customerid, date))
```

For Oracle, you need to define an additional variable to capture the new id:

```
#   Oracle
id = cursor.var(oracledb.NUMBER)
cursor.execute(insertsale, (customer, date, id))
```

To retrieve the new sale id, that depends on whether the id is returned from the INSERT statement or not.

For those PostgreSQL and MSSQL, which return a value, you can fetch that value using

```
#   PostgreSQL, MSSQL
saleid = cursor.fetchone()[0]
```

The `fetchone()` method returns the first (and subsequent row from the result set) as a tuple. Here, we want the first and only item.

For SQLite and MariaDB/MySQL, which don't return a value, there is a special `lastrowid` property:

```
#   SQLite, MariaDB/MySQL
saleid = cursor.lastrowid
```

For Oracle, the new sale id is sort of in the `id` variable, but you still need to extract it completely:

```
#   Oracle
saleid = int(id.getvalue()[0])
```

Here's the code so far:

```python
def addsale(customerid, items, date):
    insertsale = '...'         #    Add new sale
    insertitems = '...'        #    Add sale items with books
    updateitems = '...'        #    Update sale items with book prices
    updatesale = '...'         #    Update sale with total

#    PostgreSQL, MSSQL
    cursor.execute(insertsale, (customer, date))
    saleid = cursor.fetchone()[0]

#    SQLite, MariaDB/MySQL
    cursor.execute(insertsale, (customer, date))
    saleid = cursor.lastrowid

#    Oracle
    id = cursor.var(oracledb.NUMBER)
    cursor.execute(insertsale, (customer, date, id))
    saleid = int(id.getvalue()[0])

    return saleid
```

Remember not to mess around with indentation. All of the code should be one level in to be part of the `addsale()` function.

Adding the Sale Items

The sale items are a collection. Unlike SQL, Python thrives on variables, and, in particular, Python loves collection variables.

For the multiple items, you can use a tuple or a list. Here, we'll use a tuple simply to highlight the fact that we're not going to change the values. Tuples are immutable.

Each item will consist of two parts: the book id and the quantity. We could have used tuples for that too, but then the parts would be anonymous. In a future, more complex project, it would make it harder to maintain. Instead, we'll use dictionary objects.

The sale items would look like this:

```
(
    { 'bookid': 123, 'quantity': 3},
    { 'bookid': 456, 'quantity': 1},
    { 'bookid': 789, 'quantity': 2},
)
```

Within the function, the tuple will appear in the `items` variable. We can iterate through the tuple using the `for` loop.

In each iteration, we'll execute the `insertitems` statement, which inserts one item at a time. The data will be a tuple with the sale id from the previous step, as well as the `bookid` and `quantity` members of the dictionary object.

The code will look like this:

```
for item in items:
    cursor.execute(insertitems, (saleid, item['bookid'],
        item['quantity']))
```

Our function so far will resemble this:

```
def addsale(customerid, items, date):
    #   SQL Strings

    #   cursor.execute
    #   saleid

    for item in items:
        cursor.execute(
            insertitems,
            (saleid, item['bookid'], item['quantity'])
        )

    return saleid
```

The rest is easy.

Completing the Sale

We now need to run the two SQL statements to update the sale items and update the sale:

```
cursor.execute(updateitems, (saleid,))
cursor.execute(updatesale, (saleid, saleid))
connection.commit()
```

The updateitems query needs only the sale id. Even though it's only one value, it still needs to be in a tuple, which is why there's the extra comma at the end. The updatesale query needs the sale id twice, once for the main query and once for its subquery.

At the end of the job, you need to commit the transaction, which means to save the changes permanently in the database. Otherwise, the whole process is a waste of time.

The function now looks like this:

```
def addsale(customerid, items, date):
    #   SQL Strings

    #   cursor.execute
    #   saleid

    for item in items:
        cursor.execute(
            insertitems,
            (saleid, item['bookid'], item['quantity'])
        )

    cursor.execute(updateitems, (saleid,))
    cursor.execute(updatesale, (saleid, saleid))
    connection.commit()

    return saleid
```

Now you can try it out. You'll need

- The customer id (42)

- The sale items as a tuple of dictionary objects, as before

- The current date and time

To get the current date and time, you'll need to import from the `datetime` module; you can then use the `.now()` method:

```
from datetime import datetime
print(datetime.now())
```

The completed script will resemble this:

```
from datetime import datetime
import ...                          #   import connection module
connection = ... . connect(...)     #   connect to database
cursor = connection.cursor()     #   get cursor object

def addsale(customerid, items, date):
    ...

addsale (
    42,                                      #   customer id
    (                                        #   items
        { 'bookid': 123, 'quantity': 3},
        { 'bookid': 456, 'quantity': 1},
        { 'bookid': 789, 'quantity': 2},
    ),
    datetime.now()                           #   current date/time
)
```

You now have a reusable function to add sales.

Index

A

Aggregate filters, 171–173, 210
Aggregate functions, 19, 275
   basic functions, 163, 209
   contexts, 163
   count(*) OVER (), 279
   CTE, 281, 282
   daily totals *vs.* grand totals, 282
   day-by-day summary, 282
   day number, 281
   DBMS, 164, 417
   descriptions, 166
   each day sales, 280, 281
   NULL, 166
   numerical statistics, 165
   OVER (), 284
   percentage symbol, 283
   sales totals, 279, 280
   strings and
      dates, 165
   total/sum(total) OVER(), 282
   weekday, percentage/sorting, 283
Aggregate queries, 18–19, 92, 101, 168,
      246, 252, 266, 275, 405
Aggregate window functions
   daily sales view, 287
   framing clause, 285–287
   ORDER BY clause, 284, 285
   sliding window
      daily totals, 288, 289
      dates, 289
      framing clauses, 288

      sliding averages, 289
      week averages, 290
Aggregating data
   aggregate filter, 171–173
   calculated values
      arbitrary strings, 181, 183
      CASE statements, 177, 178
      CTE, 176
      customers, 173, 174
      delivery statistics, 179–181
      GROUP BY clause, 173, 176
      month name, 175
      monthnumber, 174
   clause order, 169, 170
   distinct values, 170
   error message, 168
   FROM/WHERE clauses, 169
   GROUP BY () clause, 167, 168
   group concatenation, 183, 185
   grouping sets, 195
      CUBE, 195
      data, 186, 187, 189
      GROUP BY clause, 185, 186
      renaming values, Oracle, 199–201
      ROLLUP, 196, 197
      sorting results, 197–199
      totals, 185
   query, 167
   subtotals, 211
   UNION clause
      CTE, 195
      grand total, 191

© Mark Simon 2023
M. Simon, *Leveling Up with SQL*, https://doi.org/10.1007/978-1-4842-9685-1

Aggregating data (*cont.*)
    levels, 194
    query, 189
    SELECT statements, 191, 192
    sorting order, 193, 194
    state/customer ids, 189, 190
    summaries, 189, 193
   virtual table, 167
Aggregating process, 210
Aliases, 11, 16, 32, 108–110, 159, 255, 320
ALTER TABLE statements, 20, 42, 44, 55
The American Standard Code for
    Information Interchange (ASCII),
    134, 135, 160
Arithmetic mean, 201, 203

B

Books and authors, 60, 62, 64, 70–72, 86, 94
BookWorks, 1, 2, 5
Business rules, 39

C

Caching table, 234
Calculations, 106, 113, 119, 275
Calculations in SQL
   CASE expression
     CASE ... END expression, 151
     ELSE expression, 152
   with NULLs
     coalesce(), 115–117
   using aliases
     AS keyword, 108
Calculations in SQL, 106
   built-in functions, 107
   CASE expression, 160
     coalesce, 154–158

     short-circuited, 152
     uses of CASE, 152–153
   casting, 122–127
   coding languages, 122
   data types, 122
   date operations, 160
     date arithmetic, 148–151
     date extracting in Microsoft SQL, 145
     date extracting in PostgreSQL,
      MariaDB/MySQL and
      Oracle, 144–145
     date/time, entering and
      storing, 140–141
     formatting a date, 146–147
     getting, date/time, 142
     grouping and sorting, date/
      time, 143–144
   forms, 158
   FROM clause, 121
   individual/multiple columns, 106
   with NULLs, 113–115
     author names, 117–118
   numeric (*see* Numeric calculations)
   ORDER BY clause, 120, 121
   SELECT clause, 118, 121
   using aliases, 108
     alias names, 109–110
     AS is optional, 110–111
     basic SQL clauses, 112–113
   WHERE clause, 119
cast() function, 17, 122–127, 130, 159
Casting types, 159
CHAR(length), 132
CHECK constraint, 20, 39, 43–44, 47, 55
ck_customers_postcode, 34
coalesce() function, 16, 40–42, 115–118,
    154, 159
Collation, 133, 134, 160

Columns, 28
    changing the town, 36–37
    countries table, 36–37
    CREATE VIEW statement, 33
    DROP COLUMN, 34
    foreign key, 29
    old address columns, 32
    primary key, 29
    SELECT statement, 32
    street address column, 38
    UPDATE statement, 31
Common Table Expressions (CTEs),
      267, 268
    aggregates
        duplicate names, 319, 320
        most recent sale, per
          customer, 317–319
    benefits, 268
    calculations
        monthly totals, 271, 272
        price groups, 269
        query, 270
        sales table, 270, 271
        WITH clause, 270
    constants
        deriving constants, 316, 317
        hard-coded, 314
    duplicated names
        consolidated list, 324, 325
        id, 323
        info column values, 324
        layout, 323
        parameter names, 323
        phone number, 323
        query, 322
        results, 323
    FROM subquery, 364
    hard-coded constants, 315, 316
    multiple chain, 322
    multiple CTEs, 365
    nesting subqueries, 322
    parameter names, 320, 321, 365
    recursive (*see* Recursive CTEs)
    subquery, 321, 322
    syntax, 268, 269
    uses, 364
    variables, 313, 314
    virtual table, 273, 364
Computed column/calculated
      column, 231
    creation, 231
    data, 232
    DBMSs, 232
    mini-view, 231
    ordered datetime column, 231
    read-only virtual column, 231
    types, 232
    VIRTUAL, 234
Concatenation, 133, 135, 136, 160,
      416, 417
Constants, 314
    deriving constants, 316, 317
    hard-coded, 314–316
Correlated subquery, 31, 100, 239, 240,
      243, 244, 259, 272
countries.sql, 36
CREATE TABLE statement, 20
CROSS JOIN, 18, 264, 335
Cultural notes
    address
        pattern, 407
        postcodes, 408
        states, 407, 408
        towns, 407
    currencies, 409
    dates, 409

Cultural notes (*cont.*)
    email addresses, 409
    measurements, 409
    phone numbers, 408, 409
    prices, 409

D

Data, 6, 7
Database, 6, 59, 93, 94, 103
Database design, 76
Database integrity, 39
    CHECK constraint, 47
    column constraints, 48
    domain, 39
    familyname, 48
    nullable column, 40, 41
        ALTER TABLE statement, 44
        CHECK constraint, 43
        changes in SQLite, 44
        DEFAULT value, 43
        NOT NULL constraint, 41, 42
    standard constraint types, 39
    suggestions, 45, 47
    table constraint, 48
    UNSIGNED INT, 47
Database Management Software (DBMS),
        3, 261, 299
    aggregate functions, 417
    calculations
        arithmetic, 416
        concatenation, 416, 417
        date functions, 416
        formatting functions, 416
        SELECT without FROM, 415
        string functions, 417
    database client, 4
    data manipulation, 418

    filtering (WHERE)
        case sensitivity, 414
        dates, 414
        string comparisons, 414
        wildcard matching, 415
    joining tables, 417
    MariaDB, 3
    primary keys,
        autoincrementing, 418
    quirks and variations (*see* Quirks and
        variations)
    rule, 411
    sample database, 4–5
    sorting (ORDER BY), 413, 414
    table literals, 342
    table manipulation, 418
    triggers, 370
    writing code, 411
Database tables, 2, 8
    basic principles, 9
    changes to table structures, 56
    columns, 8
    customers table, 8
    improved database design, 57
    indexes (*see* Index)
    table design and columns (*see*
        Columns)
    temporary table, 9
    virtual table, 10
    well-designed table, 9
Data Definition
    Language (DDL), 368
Data Manipulation Language (DML),
    368, 370, 381
Data rules, 27, 39
Data types
    aggregate queries, 19–20
    calculating columns, 15

calculating with NULLs, 16
   aliases, 16
   CASE expression, 16
   subqueries, 16
   views, 17
date literals, 12
joins, 17
   join types, 18
   ON clause, 18
   syntax, 17
number literals, 11
string literals, 11
Date functions, 412, 416
Deciles, 305–308
Denormalized data, 56
Design principles, 54, 60
Domains, 26, 39

E

extract() function, 144

F

Foreign key, 21, 27, 29, 30, 36, 38, 39, 50,
   55, 61, 68, 81, 276, 369
Formatting functions, 130, 132, 145, 160,
   187, 416
Frequency table, 201, 203, 205, 206

G

GROUP BY clause, 19, 211
grouping() function, 198–200

H

Histograms, 201, 202

I

Index, 49, 56
   anonymous index, 51
   author's name, 51
   books table, 50
   clustered index/index organized
      table, 50
   costs, 49
   CREATE INDEX, 50
   customers table, 49
   HAVING clause, 53
   primary key, 50
   SELECT clause, 53
   UNIQUE column, 50
   Unique Index, 52, 54, 56
IN operator, 13, 238, 247, 250, 261
Information, 6, 7

J, K

Joining tables, 104, 417

L

LATERAL JOIN (CROSS APPLY)
   adding columns
      expression, 263, 264
      principle, 264
   multiple columns
      aggregate query, 266
      FROM clause, 265, 266
      list of customers, 266, 267
      results, 266
      SELECT clause, 265
   query, 262, 273
   SELECT clause, 262
   WHERE clause, 262
ltrim() function, 118, 138

M

Many-to-many relationship, 61, 103
    associated data, 92
    association, 81
    associative/bridging table, 78, 79
    book genres, 81
    book, multiple authors, 90, 91
    book table, 78
    combination, 81
    CTE/aggregate query, 92
    data, 79
    database, 90, 93
    genres table, 78
    list of books, 93
    result, 80
    sales and saleitems tables, 90
    SELECT statements, 78
    table designing, 81
    tables, 91
    table structure, 91
    UNIQUE constraint, 81
Many-to-many tables
    associative table, 83
    bookgenres table, 83
    book's id, 83
    INNER JOIN, 82
    number of rows, 82
    results, 83
MariaDB, 3, 126, 147
MariaDB/MySQL
    connection, 426
    quirks and variations, 420
    SQL strings, 435
    triggers, 374, 375
    variables, 394, 398, 400
Median, 207, 208
Microsoft quirks and
    variations, 419

Microsoft SQL, 145, 148, 224, 225, 229, 264, 270, 304
    connection, 425
    module, windows, 422–424
    recursive CTEs, 328
    SQL strings, 434, 435
    table literals, 344
    triggers, 376, 377
    variables, 313, 314, 394, 400, 402
Mode, 205–207
Multiple tables
    adding author, 95, 96
    adding book, 97
    authors table, 94, 95
    books table, 94
    child table, 104
    joins, 104
    new book, 94
    new sale
        addition, 98
        process, 98
        sale completion, 101, 102
        sales items, 98, 100
        sales table, 98, 99
    parent table, 104
    query, 104
Multiple values, 55
    bookgenres table, 77
    CTE, 84
    GROUP BY query, 84
    id and title columns, 85
    joins, 86
        authors/genres tables, 86
        bookdetails view, 87
        books table, 87
        dataset, 86
        filtered list, 89
        filtering, 88

genre details, 88

genre names, 88

genres table, 89

query, 87

side effects, 89, 90

list, 85

multiple genres, book, 76

principles, 76

SELECT statement, 84

string_agg(column,separator)
    function, 85

MySQL, 147

N

Natural key, 36

Normalization, 26, 27, 56

Normalized database, 26–28

Normalized tables

properties, 55

ntile()

cast(... AS int), 308

customers hights, 307

decile/row_decile, 308

deciles, 306

group size, 307, 308

NULL heights, 306

rank_decile/count _decile, 309

NULLs, 7, 13, 159, 210, 276

NULL strings, 113, 118, 125, 136

Numeric calculations, 127, 160

approximation
        functions, 129–130

basic arithmetic, 127–128

formatting functions, 130–132

mathematical
        functions, 128–129

string (see String calculations)

O

ON DELETE CASCADE clause, 276, 369

One-to-many relationship, 60, 103

books and authors, 62–64

books and authors view, 70–72

child table/parent table, 62

JOIN, 62, 63

NOT IN(...), 69, 70

one-to-many joins

books and authors, 64, 65

combinations, 69

FULL JOIN, 67

INNER JOIN, 65–68

LEFT JOIN, 65, 67

NOT NULL, 67

NULL, 68

options, 65

OUTER JOIN, 66

rows, 69

subquery, 68

unmatched parents, 68

Oracle, 63

uses, 61

One-to-many tables, 93

One-to-maybe relationships

contradiction, 73

customers table, 75

customers, 74

join, 73, 74

LEFT JOIN, 75

secondary table, 73

SELECT *, 75

vip table, 73, 74

VIP columns, 75

One-to-one relationship, 61, 72, 103

Oracle, 147, 229, 304

connection, 427

quirks and variations, 420

Oracle (*cont.*)
  SQL strings, 436
  triggers, 378–380
  variables, 394, 402–404
ORDER BY clause, 12, 14, 22, 120, 121, 158,
    201, 211, 219, 220, 238, 272, 284, 293

P

Pentiles, 305
percentile_cont() function, 207
Percentiles, 305
Pivoting data
  aggregate query, 387
  customerdetails view, 384
  customerinfo CTE, 385
  database tables, 382
  definition, 382
  design, 382
  general rule, 382
  grouping, 386
  layout, 381
  ledger table, 381
  multiple CTEs, 384
  pivot feature, 389, 390
  purpose, 383
  separate totals, 388
  spreadsheet program, 382
  status CTE, 385
  testing, 386, 387
  total sales, 383
  UNPIVOT feature, 391–393
Pivot tables, 365, 367, 383
  advantages, 382
  creation, 383, 405
  definition, 405, 406
  MSSQL/Oracle, 405
  raw table data, 405

Planned relationships, 61
PostgreSQL, 147, 222, 223, 229, 264
  connection, 426
  quirks and variations, 419
  SQL strings, 433
  triggers, 373, 374
  variables, 394, 397, 398
Previous and next rows
  comparing sales, 310
  daily sales, 309
  lag and lead, 309, 311
  missing dates, 311
  OVER clause, 309
Python, 421
  connection, 421
    cursor, 424
    mysql.connector, 426
    oracledb, 427
    psycopg2, 426
    pyodbc, 425
    sqlite3, 424
  database connector module
    exceptions, 422
    shell/command line, 422
  fetching database, 427, 429
  module, 421
  MSSQL module, windows, 422–424
  new sale
    addition, 437, 438
    code, 432
    completion, 440, 441
    customerid, 432
    methods, 432
    sale items, 438, 439
    SQL strings, 431, 432
    steps, 431
  parameters, query, 429–431
  SQL strings, 432

MariaDB/MySQL, 435
MSSQL, 434
Oracle, 436
PostgreSQL, 433
SQLite, 434
triple quote characters, 432

Q

Quirks and variations
MariaDB/MySQL, 420
Microsoft, 419
Oracle, 420
PostgreSQL, 419

R

Ranking functions, 275, 297
basic functions, 298
count(*), 298
customer heights, 298, 299
dense_rank(), 298
examples, 299
exceptions, 299
expressions, 300
framing clause, 297
ORDER BY value, 298
paging results
CTE, 303
OFFSET ... FETCH ... clause, 304
pricelist view, 303
prices, 305
PARTITION BY
CASE ... END expression, 301
columns, 302
expected order, 301
order date, 301, 302
row_number(), 300
rank(), 298

row_number() function, 297
Recursive CTEs
(cte(n)), 327
daily comparison, missing days
daily_sales view, 333, 334
DBMSs, 335
finding dates, 334
LEFT JOIN, 336
sequence of dates, 334, 335
vars and dates, 335
forms, 325, 326
JOIN, missing values, 331–333
parts, 326
sequence, 327
adding day, 329, 330
creation, 326, 328
dates, 329, 331
MSSQL, 328, 331
series of number/dates, 331
WHERE clause, 328, 331
traversing hierarchy
cleaner result, 341, 342
employees table, 337
multilevel, 338–341
single-level, 337, 338
supervisorid column, 337
supervisor's name, 337
uses, 327, 365
Relational model, 6
Relationships, 102
planned, 61
types, 60, 103
unplanned, 61

S

sales table, 98, 99, 102, 155, 186, 270, 276, 280, 318
Scalar function, 221

SELECT statement, 12, 234, 235
Single value query, 237
SQL, 394
    basic SQL, 10–11
    data types, 11–12, 412
    dates, 412, 413
    feature, 102
    query, 368
    quotes, 413
    semicolon, 412
    writing, 10, 411
SQL clauses
    clause order, 12
    limiting results, 14–15
    multiple assertions, 13
    ORDER BY clause, 14
    SELECT clause, 12
    sort strings, 15
    WHERE clause, 13
    wildcard patterns, 13
SQLite, 64, 140, 261, 395, 406
    connection, 424
    SQL strings, 434
    triggers, 378
Standard deviation, 208, 209
Statistics, 212
String calculations, 160
    ASCII and Unicode, 134–135
    case sensitivity, 134–135
    CHAR(length), 132
    concatenation, 135–136
    data types for strings, 132
    string functions, 137–139
    VARCHAR(length), 132, 133
String functions, 417
Subqueries
    column names, 243
    complex query, 239
    correlated, 239, 240, 242, 243, 245, 272
    cost, 239
    definition, 238
    expression, 245
    FROM clause, 272
        NULL, 255
        price groups, books, 253
        SELECT statement, 254, 255
        summarizing table, 253
    GROUP BY clause, 254
    IN() expression, 273
    nested subqueries, 255–257
    non-correlated, 239, 240, 242, 272
    ORDER BY clause, 272
    SELECT clause, 242
        aggregate query, 245
        correlated subquery, 243, 244
        join, 243
        non-correlated subquery, 244
        window functions, 244
    uses, 238, 272
    WHERE clause, 242
        aggregates, 246
        big spenders, 246–249
        duplicate customers, 251, 252
        last order, 249–251
    WHERE EXISTS (...), 258, 273
        correlated subquery, 259, 260
        FROM dual, 258
        IN() expression, 260, 261
        non-correlated subquery, 259
        SELECT NULL/SELECT 1/0, 259
        testing, 258

T

Table, 213
Table design, 20

constraints, 20
data manipulation statements, 21
foreign key, 21
indexes, 21
set operations, 22–23
types of data, 20
Table literals
   data
      anchor member, 360, 361
      CTE, 359
      recursive member, 361–364
   DBMSs, 343
   definition, 342
   lookup table, 352, 353
   MSSQL, 344
   sorting
      advantage, 351
      data CTE, 349
      names, 349
      sales per weekday, 348
      sequence number, 350, 351
      strings, 348
      summary CTE, 349
   standard notation, 343
   statement, 342
   string
      anchor member, 355, 356
      recursive CTE, 353–355
      recursive member, 356
      rest, 355–358
      splitting, 354
      WHERE rest<>, 358
   testing
      age calculation, 344, 347, 348
      dates CTE, 345, 346
      series of dates, 345
   virtual table, 342
Table Valued Function (TVF), 221

DBMSs, 221
definition, 233
Microsoft SQL, 224, 225
PostgreSQL, 222, 223
pricelist(), 222
Temporary table
   benefits, 234
   creation, 228, 229
   database, 230
   INSERT ... SELECT ... statement, 229
   query, 230
   SELECT statement, 229
   TEMP, 229
   uses, 230
towns.sql., 28
Towns table, 28, 29, 31, 32, 35–37, 384
Triggers
   activity table, 368
   archive table, 370
   creation, 404
   data, archiving, 370, 372
   data deletion, 369
   DBMSs, 370, 405
   definition, 367, 368, 404
   deleted_sales, 370
   foreign key, 369
   logging table, 368
   Logon triggers, 368
   MariaDB/MySQL, 374, 375
   MSSQL, 376, 377
   NULL sales, 369
   Oracle, 378–380
   PostgreSQL, 373, 374
   pros and cons, 380
   rental table, 368
   sales table, 369
   sales deletion, 372
   SQLite, 378

Triggers (*cont.*)
    syntax, 369
    types, 368
    uses, 368, 406
Triggers, 365

U

Unicode, 134, 135
UNIQUE clause, 29
Unplanned relationships, 61

V

Value, 7
Value functions, 275
VARCHAR(length), 132, 133
Variables, 313, 314
    code blocks, 395, 396
    DBMSs, 394, 406
    definition, 367, 394
    function/procedure, 394
    MariaDB/MySQL, 394, 398, 400
    MSSQL, 394, 400, 402
    Oracle, 395, 402–404
    PostgreSQL, 394, 397, 398
    purposes, 406
    statements, 405
    system variables, 394
    uses, 406
Variables, 315
Views, 55, 214
    aupricelist, 218, 219
    benefits, 215
    caching data, 227
    cascade views, 220
    conditions, 217
    convenience, 225
    CREATE VIEW ... AS clause, 217

DBMS, 215
    external applications, 227
    importance, 215
    interface, 225, 226
    limitations, 215
    materialized views, 228
    ORDER BY clause, 219–221
    pricelist view, 216
    SELECT *, 220
    SELECT statement, 233
    syntax, 214
    TVF (*see* Table Valued Function (TVF))
    uses, 214
Virtual tables, 213, 214, 233, 235
    multiple rows and multiple columns, 236
    one column and multiple rows, 236
    one row and one colum, 235, 236
    query, 237

W, X, Y, Z

WHERE clause, 13, 17, 20, 100, 112, 113,
        119, 210, 242, 258, 262, 328
Window clauses, 312
Window functions, 236, 244, 274–276, 311
    aggregate windows, 277–279
    ORDER BY clause, 277
    OVER () clause, 276, 277
    PARTITION BY clause, 276
    subtotals
        expressions, 290
        monthly totals, 291
        ordered_month, 293
        PARTITION BY multiple
            columns, 294–296
        PARTITION BY/ORDER BY, 292, 293
    syntax, 276
    window, 276

Printed in the United States
by Baker & Taylor Publisher Services

Printed in the United States
by Baker & Taylor Publisher Services